795
NEW

Celadon Wares

A. Dish with carved peony decoration. Chinese. Northern Celadon ware. Sung dynasty. Diameter 8.1 in. Fitzwilliam Museum, Cambridge.

CELADON WARES

G. St. G. M. Gompertz

FREDERICK A. PRAEGER, Publishers
New York · Washington

BOOKS THAT MATTER

Published in the United States of America in 1969
by Frederick A. Praeger, Inc., Publishers
111 Fourth Avenue, New York, N.Y. 10003

© 1968, in London, England, by G. St. G. M. Gompertz

Library of Congress Catalog Card Number: 68-19135

Printed in Great Britain

Contents

Illustrations

COLOUR PLATES

MONOCHROME PLATES
(at the end of the Book)

9

9. Elliptical brush washer (under side). Ju ware. Sung dynasty.

10. Ewer with strongly carved floral decoration under a pale green glaze. Possibly Tung ware. Sung dynasty.

11. Dish with carved peony decoration. Northern Celadon. Sung dynasty.

12. Octagonal bottle with long tapering neck ending in a flange; horizontal ribs on the neck and body. Greyish blue glaze. Southern Kuan ware. Sung dynasty.

13. Dish in the form of an eight-petalled flower covered with a bluish glaze. Southern Kuan ware. Sung dynasty.

14A. Bottle with globular body and long neck. Bluish green glaze. Southern Kuan ware (?). Sung dynasty.

14B. Brush-washer with straight sides and wide flat rim. Bluish green glaze. Southern Kuan ware (?). Sung dynasty.

15. Dish with pair of fishes in relief. Lung-ch'üan ware. Sung dynasty.

16. Mallet-shaped vase with phoenix handles. Lung-ch'üan ware (Kinuta type.) Sung dynasty.

17A. Vase with flaring mouth and floral decoration in relief. Lung-ch'üan ware. Sung dynasty.

17B. Bowl with moulded petal decoration outside. Lung-ch'üan ware. Sung dynasty.

18. Vase of spotted celadon ('tobi seiji'). Lung-ch'üan ware. Sung dynasty.

19. Basin with dragon, cloud scrolls and flowers in biscuit relief. Lung-ch'üan ware. Yüan dynasty.

20. Beaker-shaped vase in the form of a bronze 'tsun'. Lung-ch'üan ware. Ming dynasty.

21. Dish with incised floral design; on the base the inscription 'Great Ming Hsüan-te Lung-ch'üan'. Lung-ch'üan ware. Ming dynasty.

22. Incense burner with cover surmounted by a lion. Korean ware Koryŏ dynasty.

ILLUSTRATIONS

35. Globular bottle with two small loop-handles. Sawankalok ware.
36A. Small Sawankalok Celadon bottle with ears.
36B. Modern Celadon wares produced by the Thai Celadon Company of Bangkok.

SKETCH MAPS

Introduction

The name celadon has been given to ceramic wares of a distinctive green or bluish green colour. It was first applied to the Chinese wares made at the Lung-ch'üan potteries in Chekiang during the Sung and Ming periods, which seem to have become known in Europe during the late sixteenth or seventeenth century. However, although it has come into use to describe similar porcellanous wares made in China, Korea and Siam, there has been a conventional limitation in its use, and it has not as a rule been applied to the official, or imperial, Chinese wares, although these often have a very close resemblance in colour and tone to the Lung-ch'üan or other celadons. While it is a descriptive, non-scientific term, this restriction in its use is quite irrational, and it will here be applied impartially to all wares which come within its fairly well understood colour range.

Celadon ware has been made in China for about two thousand years, and it has always been greatly admired in the Far East and, indeed, throughout Asia. It was exported in vast quantities from the ninth century at least until the fourteenth, and relics of this trade have been found in many Eastern countries, especially in India, Persia, Turkey, Egypt and throughout Eastern and South-eastern Asia. It was a late arrival in Europe but has been in great demand by Western collectors and connoisseurs during the present century, and specially fine examples now command prices as high as £8,000. However, there is no reason for those who develop an interest in one of the loveliest types of Chinese ceramics to take fright at the astronomical sums paid for the finest specimens. It is quite possible to acquire less superlative examples for sums of £30 to £60, while some very delightful small pieces are often obtainable for a few pounds. A new collector

of limited means would be well advised to begin by concentrat-
ing on the less expensive but also attractive Ming period celadons,
which can be most effectively displayed on furniture with a
mahogany veneer—the ruddier the tone of the mahogany, the
lovelier is the contrast with the celadon green. Later, if these Ming
wares are found to lose some of their appeal, it will always be
possible to switch to the more highly valued Sung wares, and
this is most unlikely to result in financial loss—apart from the
higher prices which will have to be paid for the earlier wares.

The interest in Chinese and other oriental ceramic wares has
greatly increased in recent years and has been stimulated by the
prodigious amount of archaeological work which has been undertaken by the Chinese authorities. Up to the Second World War
the major discoveries had all been made by Japanese and Western
students, but these were relatively few in number, though all of
great importance in view of the ignorance then prevailing; but
now hardly a month goes by without further significant finds
being reported in one or other of the Chinese archaeological
journals. The writer of this book has made a special effort, first, to
become familiar with all available details of the pre-war Japanese
studies—which has been facilitated by his long sojourn in Japan and
close association with Fujio Koyama and other Japanese experts—
and, second, to keep abreast of the field-work which has been in
progress during the past ten or fifteen years in China; his objective
being to collate Western and oriental views and reach a synthesis
of opinion on an authoritative, but by no means a definitive,
basis. For there is still a vast amount to be learned and much that
remains uncertain, speculative or even completely obscure—and
herein lies one of the chief attractions of the subject. In no sphere
is such a synthesis of greater importance than that of the celadon
wares, which Koyama has called the 'backbone' of oriental
ceramics. However, the writer's special interest and enthusiasm
have always been aroused by the Korean celadons, which are still
so little known in the West and constitute one of the major
achievements in oriental ceramic art. A four-week stay in Siam has

been of great assistance also in acquiring more familiarity with Sawankalok celadon, which is qualitatively inferior to the best Chinese and Korean wares, yet none the less possesses a distinctive charm of its own.

Thanks are due to the museums and collectors who have graciously permitted reproduction of some of their greatest treasures. It is hoped that this small handbook will stimulate further interest in a subject which many devotees have found of never-ending fascination. So far as known, it is the first book ever published which deals with all the varieties of celadon ware, Chinese, Korean, Japanese, Siamese and modern.

A short Bibliography of books in English, also the principal recent works in Japanese and Chinese, will be found at the end of this study, and references to specialist articles in Western and Far Eastern journals are given in footnotes, so that readers will be able to follow up any particular lines in which they may be interested.

CHINESE DYNASTIES & PERIODS

Shang-Yin	*c.*1600–*c.*1030 B.C.
Chou	*c.*1030–249 B.C.
(Warring States *c.* 481–221)	
Ch'in	221–206 B.C.
Han	206 B.C.–A.D. 220
(Former Han 206 B.C.–A.D. 25)	
(Later Han A.D. 25–220)	
Six Dynasties	220–589
(Three Kingdoms, Northern &	
Southern Dynasties)	
Sui	581–618
T'ang	618–906
Five Dynasties	907–960
(Wu-Yüeh 907–978)	
Sung	960–1279
(Northern Sung 960–1127)	
(Southern Sung 1127–1279)	
Yüan	1280–1368
Ming	1368–1644
Ch'ing	1644–1912

KOREAN DYNASTIES & PERIODS

Lo-lang Chinese Han Colony	108 B.C.–A.D. 313
Three Kingdoms	*c.*57 B.C.–A.D. 668
(Silla *c.*57 B.C.–A.D. 668)	
(Koguryŏ *c.*37 B.C.–A.D. 668)	
(Paekche *c.* 18 B.C.–A.D. 660)	
Great Silla	668–918
Koryŏ	918–1392
Yi	1392–1910
Japanese Rule	1910–1945

CHAPTER ONE

Celadon Wares—What They Are and How They Are Made

Celadon is the name given to a wide range of Chinese, Korean and other porcellanous wares which have a distinctive bluish green or grey-green glaze. It is essentially a colour term, applied without regard to the ingredients or method of manufacture and thus quite unscientific and equivocal. However, it has come into general use and is unlikely to be superseded by a more exact term, so that it is desirable to begin our survey by examining its derivation and then providing an acceptable definition which will make it clear what types of ware are regarded as falling under this heading.

The name celadon is believed to have been taken from that of a character, the shepherd Céladon, in one of the plays founded on the romance, *l'Astrèe*, written by a seventeenth-century French playwright, Honoré d'Urfé. The arrival of Chinese celadons in Europe is thought to have taken place about the same time, and it is supposed that they were given this name because the character in the play always appeared on the stage attired in grey-green clothing. However, a variant theory is put forward in brochures published by the Thai Celadon Company of Bangkok—the only modern factory engaged in the large-scale production of celadon ware; according to these, the name is derived from Sanskrit words meaning 'stone' and 'the act of wearing', so that the composite word is translated as 'sheathed in stone'. Another often-repeated view is that the name was derived from Saladin, Sultan of Egypt, who is said to have sent forty pieces of the ware to Nur-ed-din,

CELADON WARES

Sultan of Damascus, in the year 1171. Finally, classical scholars may recall a reference in the seventh book of Homer's *Iliad* to the Greek river of the same name: 'where Celadon rolls down his rapid tide'; and it would be a pleasing conceit to hold that the term was derived from the green waters of Arcady.

The origin of the name celadon is thus not entirely clear; and its use also has been somewhat arbitrary. It is applied *par excellence* to the famed Chinese ware made in Sung times at the Lung-ch'üan or other potteries in Chekiang Province and extended by virtue of strong similarity to the Yüeh wares developed much earlier in the same region, to the 'Northern Celadons' of Honan and Shensi and to Korean and Siamese products of the same type; but it is rarely used of the imperial Chinese wares, Kuan and Ju, although some Korean celadons are equal to Ju in quality and hardly distinguishable from it in colour or texture. This is not true of the Chinese term, *ch'ing tz'u* (Japanese: *seiji*), signifying green (or blue) porcelain, which may be considered roughly equivalent and is used indiscriminately to describe wares of many different shades. There is wisdom in the oriental view that it is a mistake to attempt a precise differentiation of a range of tones which, experience tells us, are not always seen in the same way by different people.

Scientific investigation has established that the colour of the celadon glaze is due to a small percentage of iron fired in a 'reducing atmosphere', but this only adds further complications. Thus, it is difficult to associate the blue tone of Chün ware with the term celadon, though this is now known to be attributable to iron; while the celadon glaze itself will turn brown if the firing takes place in an 'oxidizing atmosphere'. Everything considered, it seems best to adopt a broad viewpoint, accepting as celadon all the wares mentioned above with the exception of Chün; furthermore, it is in practice difficult to exclude specimens which have turned brown in the firing due to oxidation, because many of these are closely bound up with the historical development of celadon ware. The potters were ignorant of the chemistry involved in the production of high-fired pottery with feldspathic glazes and were accordingly compelled to rely on empirical methods, which resulted in numerous failures. Even as late as the end of the twelfth century we find a famous Korean scholar commenting that only one in ten celadon bowls reached the desired standard.[1]

To the connoisseur celadon wares are among the chief glories of oriental ceramics: he never tires of examining each piece and noting the subtleties of tone and shade, the beauty of the carved or impressed decoration; but those without previous experience are often bewildered by the quiet, subdued coloration, which at first seems dull and monotonous. Accustomed to the variety and bright colours of European porcelain, they find it difficult to adjust their taste to a class of ware which has little immediate appeal and, like good wine, requires experience and the development of a sensitive perception to appreciate the full flavour. However, the matter should not be regarded as too recondite or precious for general recognition. It is only necessary to suspend preconceived ideas about porcelain and to return after an interval to the celadons which have attracted—or repelled—at first sight. The eye soon accommodates to a new experience and begins to take note of features which were not at first apparent; and finally, after a period of self-tuition which may be quite short or else more prolonged according to circumstance, it will often be found that one's appreciation has grown beyond expectation and that a new realm of beauty has unfolded. There are, indeed, few more lovely sights in the whole field of ceramics than a well-selected group of celadons, with their smooth, soft glaze texture and restful tones ranging from an ethereal bluish green to olive or dove-grey.

When we come to consider the history of Chinese and other oriental ceramics, we find that celadon wares have been on a pinnacle of fame from very early times, long before they were known to the West. Many of the earliest celadons, known by the generic name of Yüeh ware, were exported from South China to India, Persia and Egypt in the T'ang period (A.D. 618—906), while the

[1] Yi Kyu-bo (1168-1241) in *Tongguk Yi Sangguk-chip*, Bk. VIII. (See p. 99a. of facsimile edition produced by photo-offset, Seoul, 1959.)

celebrated Lung-ch'üan wares made later in the same region were mass-produced during the Sung and Ming periods (960–1644) and shipped in huge quantities to all parts of the Asiatic continent from Japan to Turkey, so that some thirteen hundred specimens have been preserved to this day in the Topkapu Palace at Istanbul. The mysterious beauty of the celadon glaze resulted in legendary properties being ascribed to it in Japan and to the belief that celadon table-wares would disclose the presence of poison in food, which was widespread in the Near East and even in Europe. The first celadons to arrive in Europe probably came from Egypt and India; but during the present century most of them have been imported from China or Japan; and as the number of specimens preserved in collections or excavated from tombs has been inadequate to meet the demand, the ancient kiln sites have been subjected to ruthless exploitation: hundreds of wasters, or discards, have been recovered and so far as possible restored to grace the show-cases of collectors and museums. Indeed, some of these wasters have, by a strange chance, possessed glazes of quite exceptional beauty: thus a small 'bubble-bowl' formerly owned by the late A. L. Hetherington, which had been provided with an ingenious ornamental stand to mask the blemish of a double rim, was of such an ethereal blue-green colour that it was in great demand when it was offered for sale.

It was Hetherington who first investigated the scientific principles underlying the chief glaze effects produced by the Chinese potters. His study is still one of the basic works on oriental ceramics and should be consulted by all students and collectors who wish to be informed about the chemistry of glazes.[1] An absorbing chapter entitled 'The Iron Story' deals with the astonishing range of colours achieved by use of the two chief oxides of iron, ferric oxide and ferrous oxide, from the blue-green, grey-green and emerald of celadon ware to the reds, browns and blacks of the *temmoku* family, which often appear in spots or splashes

[1] A. L. Hetherington, *Chinese Ceramic Glazes*, Cambridge University Press, 1937.

when an excess of oxide comes to the surface during cooling of the saturated solution. Some years earlier Hetherington had drawn attention to the beauty of celadon ware when used in modern interior decoration for various practical purposes as an adjunct to old furniture,[1] and he was to introduce an exhibition of celadon wares with the following passage:

'There is probably no section of Chinese ceramics more generally popular or sought after than celadon, and the appeal made by the closely related Korean wares is not dissimilar. The restful colour of the glaze, ranging from a bluish green through varying shades of greyish green to a sea-green and a deep olive-green harmonizes with furniture; and many of the utensils such as bowls, dishes, jardinières of varying sizes and shapes lend themselves well for the purpose they were constructed to serve. Moreover, being stoutly potted, these utilitarian vessels have withstood the ravages of time and have remained available in greater numbers relatively than the more delicate porcelains of much later date.'[2]

The softness of colour and texture of celadon ware is one of its most prized qualities and is due to the scattering of light from innumerable particles of undissolved material and minute bubbles held suspended in the glaze. This results in an opacity which produces a strong resemblance to jade, and since jade was regarded by the Chinese as the most precious of all materials, it is safe to say that this must have been a prime objective, one which caused celadon wares to possess a special value in the eyes of the Chinese connoisseur. Indeed, the leading Japanese authority, Fujio Koyama, has with good reason claimed that celadon wares are the 'backbone' of oriental ceramics. He maintains that celadons are more native, or inherently Chinese, than other kinds of porcellanous wares which developed in the same period but all owed

[1] A. L. Hetherington, 'Celadon Porcelain: its Story and Decorative Value', *Old Furniture Magazine*, April, 1928.

[2] *Catalogue of an Exhibition held by the Oriental Ceramic Society, Celadon Wares*, London, 1947.

something to external influence. Moreover, the history of porcelain in China is closely connected with celadon ware. The precursors of celadon ware were the so-called 'proto-porcelains' which marked an important stage in the development of porcelain, and the earliest true porcelains by oriental definition—which accounts hardness and resonance of more significance than whiteness or translucence—were the Yüeh celadons of the T'ang and Five Dynasties periods.

The colour of celadon ware, basically due to iron in a ferrous condition, is dependent on the composition of the clay and glaze, but the most important factor is the firing in the kiln, which must attain a sufficiently high temperature and generate a reducing atmosphere to result in the blue-green or emerald tones which excite so much admiration. This means that combustion must be incomplete or smoky, when the carbon present captures the available oxygen and reduces the oxides to their metallic forms. Conditions of complete combustion, on the other hand, result in the burning gases being amply supplied with oxygen and cause the metals in the clay and glaze to produce their oxide colours: in the case of celadon this effects a change to brown or yellow. This may be apparent on one side or only part of a vessel, or it may suffuse the whole with an even brown tone known to the Chinese as *mi sê*, the colour of roasted rice, often called 'straw-coloured' or 'millet-coloured' in the West. This effect has a mellow charm of its own and may sometimes have been deliberately contrived on account of its special appeal; but in the majority of cases it is likely to have been the result of faulty technique—an inability to maintain a reducing atmosphere during the firing, at least until the glaze has reached maturity. Even a blast of air from a spy-hole at the wrong moment will nullify previous precautions, causing the green colour to disappear and leaving a sickly yellow spot on one side of a piece where it has been oxidized. Reduction should begin at about 1100°C. and continue to the end. Soft greys are said to be obtained by starting the reduction period earlier and blue by finishing the last half-hour with a

'neutral' flame. In this connection Koyama has stressed the significance of the use of wood-fuel, common to the kilns of central
and southern China—also of Korea—where celadon ware was
first developed and reached its ultimate perfection. In the north
the use of coal was more general, with the result that the Northern Celadons are mostly characterized by the olive-green tone
produced by partial oxidation. The long, smoky flames resulting
from the use of wood-fuel were effective in developing the
requisite bluish green colour, but this was at first difficult to
achieve owing to inadequate control of the firing temperature and
conditions, and the earliest celadons mostly exhibit a brownish or
olive hue and a blotched or mottled effect due to imperfect firing.

It was indeed a remarkable achievement for the Chinese potters
of two thousand years ago to contrive and maintain the high
temperatures—around 1250°–1350°C.—necessary for the production of celadons and other porcellanous wares. Their industry
and perseverance only met with success after a lengthy period of
trial and error, for it must always be borne in mind that they were
working largely in the dark, without the aid of chemistry as we
understand it. Moreover, the quality of the celadon glaze depends
to no small extent on the preliminary grinding, which in Japan
at the present time may be continued for as long as forty hours.
In fact, as modern potters well know, there are so many things
that can go wrong with the selection and preparation of both clay
and glaze, the construction of the kiln and the stoking and burning
of the fuel that it is little short of miraculous that the early Chinese
potters, working mainly by rule of thumb, were able to produce
such matchless glazes and effects. Although the modern potter is
fully versed in the scientific background and can regulate the
temperature and conditions within the kiln by using controlled
heat with the aid of oil-fired or electric furnaces, he has been
unable to achieve the individuality and charm of the twelfth-
century celadon wares with their variegated colouring and soft,
gem-like lustre.

One reason for this, no doubt, is that the Chinese potters were using unrefined, earthy materials, and the natural impurities present imparted a special character to the completed pottery. The modern artist-potter has appreciated the significance of this factor and often experiments with various types of natural wood-ash to produce the ingredients necessary as fluxing agents in stoneware glazes. However, Western potters in general have travelled further away from the natural conception of clay towards over-refined mixtures appropriately termed pastes. The oriental potter, working with natural materials and overcoming disabilities by traditional expedients, was able to utilize fully their basic properties. As stated by that great connoisseur of pottery and folk-arts, Sōetsu Yanagi, they were aided by 'the boundless power of nature existing in fire, water, clay and glaze'. Their products possess an intrinsic beauty derived in part from the very impurities which have been banished from the refined, artificial materials found on the market to-day.

The colour of the clay is of vital importance in the production of celadon ware, since the celadon glaze itself is generally transparent or semi-transparent. The employment of pure white clay, as seen in celadons of the Ch'ing period (1644–1912) made at the great pottery town of Ching-tê-chên and in many modern celadons, produces the clearest greens but results in an almost total loss of what Koyama has called the 'depth, mystery and spiritual power' of the early wares. By comparison, these later celadons seem lifeless, their attractive colouring and graceful shapes failing to make good the loss of more profound qualities. The darker clays of the early stonewares combine with the glaze in imparting a feeling of vitality, and the subtle tones and textures are partly attributable to this happy marriage of materials.

The basic components of the celadon glaze are lime, potash, feldspar, silica and iron. These form an alkaline compound necessary for the production of green from iron oxide. The fluidity or viscosity of the glaze is determined by the ratio of the alkalies, lime and potash, to the feldspar and silica. Variations in colour are often due to the thickness of the application of glaze, which in a thin coating may be almost completely transparent. An infinitesimal increase in the depth and thickness of the glaze results in intensification of the colour and an increase in opacity. The percentage of iron contained in the glaze ranges from one-half of one per cent, producing only a faint tinge, to three per cent, which results in a deep, soft green. The body clay likewise often contains iron, so that the materials were consonant and the final result owed much to their judicious blending by potters who had no scientific knowledge but relied on an intuitive flair and understanding, an ability to match means to ends and to work in full conformity with natural processes.

NOTE: Since the above was written, the author has received from Mr. Manjong Kim, Director of the Research Institute of Mining and Metallurgy, Seoul, an interesting report,[1] part of which deals with the production of the celadon glaze and its discolouration by oxidation. This shows that ferrous iron formed by reduction is highly unstable at high temperatures: 'it will oxidize quickly even by a single flash of oxygen in the kiln'. The report goes on to state that reduction may be delayed as late as 700°C., prior to which an oxidizing atmosphere is desirable 'to burn off all the organic matter contained in the raw material'. After this, however, a reducing atmosphere is essential and must be maintained until the wares have cooled down to about 300°C. to prevent re-oxidation. It had previously been thought that oxidation could never occur once an impervious glaze surface had been formed, but this was shown not to be the case: 'it was observed that oxidation takes place quickly between the liquid and gas and that oxygen in glaze containing iron oxide diffuses rapidly into the reduced glaze through fluid convection . . . just like a drop of ink in a glass of water'. Whereas 'a common green colour' results from slightly incomplete reduction during firing or slight re-oxidation during cooling, the bluish tone is extremely difficult to produce—probably only a single piece out of several hundred, 'by miraculous good fortune'. Mr. Kim adds: 'the true art appears to stand higher than ordinary eyes, because no modern reproduction could satisfy the Japanese celadon experts who visited Korea last summer'.

[1] *Report on the Development Research Plan for the Utilization of the Korean Anthracite for the Production of Refractories and Related Products*, 1967.

CHAPTER TWO

The Beginnings of Celadon
Ware in China

The origin of celadon ware is still very obscure. The earliest of all glazing effects on stoneware is seen principally round the shoulders of jars and vases and was the accidental result of wood-ash from the fuel used in the furnace falling onto the heated pottery; but it was not long before the potters made use of this fortuitous occurrence by introducing material which would supplement the process and produce a more even and consistent glaze covering the entire body of the vessels. This was facilitated by such devices as spraying or applying the glaze material with a brush, or by dipping the whole vessel into the mixture.

Early low-temperature glazes composed of lead, borax or soda may originally have come from the Near East; but they proliferated in China and were used with great effect to decorate some of the most beautiful T'ang period pottery in lovely shades of green, blue and yellow. The glaze was commonly applied to the upper parts of the vessels and ran down in waves, terminating near the base in thick drops or globules. However, side by side with these 'soft' glazes, which melted between 750° and 1000°C., the Chinese were developing stoneware with authentic though primitive feldspathic glazes, for they realized the advantages to be obtained by employing a harder and less porous kind of pottery. The first stoneware glazes were composed of a single natural stone or clay, sometimes with an admixture of a small amount of wood-ash, and were brownish in colour. The main source of the colour was iron, but the proportions were variable, since at this stage it was not artificially introduced but existed as an impurity in one or more of the natural materials used. The production of a reducing atmosphere in the kiln likewise was a gradual process, assisted by the employment of wood-fuel, and was not successfully achieved until the potters had learned by experience just how they could gain mastery over the technique.

Most of our present knowledge of the earlier celadon wares has been derived from chance discoveries and uncontrolled excavation and in many instances specimens have come onto the market without any reliable indication as to how or where they were found. However, a great number are known to have been recovered

LEADING TOWNS AND KILN SITES
IN CHINA AND KOREA

from old graves in Chekiang Province, and there is strong presumptive evidence for holding that celadon ware originated in that part of China, although an increasing number of finds are now being made north of the Yangtze and far in the interior. Some of the most primitive examples known are contained in the Ingram Collection at the Ashmolean Museum in Oxford and are believed to be precursors of the celebrated Yüeh ware made several centuries later in the same region. There is no certain means of dating these specimens, but they have a strong resemblance to various types of ritual bronzes made in the late Chou period, for which they were in all probability cheap substitutes. In other words, they were *ming ch'i*, or objects specially made for interment with the dead. This has prompted the view that they were manufactured soon after the Chou period, or in the period of the Warring States—perhaps in the third century B.C. However, Willetts has pointed out that 'bronze forms continued to be expressed in pottery long after they had ceased to be made in metal',[1] while Ayers has drawn attention to the recent discovery of similar vessels in Han tombs at Li-chu, near Shao-hsing.[2] It is thus possible that these wares, although they seem even more archaic, may actually date from the Han period—somewhere between 200 B.C. and A.D. 200. Besides their crude, bronze-like forms, they are characterized by vitreous glazes unevenly spread over the surface, imperfectly fired and often brownish yellow in colour, 'like dried gum'. There is every reason to accept the view propounded by Karlbeck and Hochstadter that they represent the ancestral type of ware from which the Yüeh celadons were developed.[3]

The next great landmark in the history of celadon ware was

[1] William Willetts, *Foundations of Chinese Art*, London, 1965, p. 261.
[2] John Ayers, *The Seligman Collection of Oriental Art*, Vol. II: *Chinese and Korean Pottery and Porcelain*, London. 1964, pp. 4–5.
[3] Orvar Karlbeck, 'Early Yüeh Ware', *Oriental Art* (Old Series), Vol. 2 No. 1, Summer 1949; also Walter Hochstadter, 'Pottery and Stonewares of Shang, Chou and Han', *Bulletin of the Museum of Far Eastern Antiquities*, No. 24, 1952, p. 100.

provided by some excavations carried out as long ago as 1924 by C. W. Bishop on behalf of the Chinese Government at Hsin-yang in southern Honan. A brick bearing a *nien hao* corresponding to the year A.D. 99 was found in the vicinity of the graves excavated and was of exactly the same type as those used for their construction, so that we have a reliable *terminus ad quem* for dating the vessels recovered, which were at first taken to be Sung celadons and included several basins decorated with 'diamond diaper' bands. This 'Hsin-yang pottery' was formerly on view at the Historical Museum in Peking and was examined by such authorities as Fujio Koyama, Sueji Umehara and Dr. Wan-li Ch'ên. Koyama described the body of the ware as 'light grey, hard, fine-grained and semi-porcellanous' and the glaze as 'semi-transparent and light olive in colour', while Bishop termed the glaze 'a pale greenish hue'. All authorities agreed, however, that the Hsin-yang pottery is datable to the Later Han period (A.D. 25–220) and is similar to Yüeh ware made in the Six Dynasties period, from A.D. 220 onwards. The importance of this discovery was that it established beyond reasonable doubt the origin of Yüeh ware— and therefore of true celadon—some time in the Han period, as first put forward with supporting evidence a few years later by the young English scholar, Brankston. After visiting one of the earliest Yüeh kiln centres at Chiu-yen, near Shao-hsing, Brankston showed in a brief but authoritative article that there were numerous Han-style Yüeh wares, many of which were probably made at Chiu-yen, and that the period during which the Chiu-yen kilns were active could be reliably placed at the first to the sixth century of our era.[1] Apart from the Hsin-yang pottery, however, no other specimens are known to have been discovered in precisely dated Han tombs, and the forms and decoration which became popular at that time and have come to be regarded as characteristically Han continued to be used well into the subsequent Six Dynasties period.

[1] A. D. Brankston, 'Yüeh Ware of the "Nine Rocks" Kiln', *The Burlington Magazine*, Vol. LXXIII, December, 1938.

There is in consequence some difficulty in dating the numerous Yüeh celadons which have come to light as a result of the pillaging of old tombs, most of which were undoubtedly made in the Six Dynasties period up to the founding of the T'ang dynasty in A.D. 618. However, as Brankston showed, some of the vessels are so strongly redolent of Han—for example large basins and dishes similar to dated Han bronzes (Plate 2), wine-cups exactly like Han lacquer wares and incense burners of the familiar Han 'hill-jar' type—that they may be assigned with a fair degree of certainty to the Han period; but the stylistic evidence is not completely trustworthy for the reason given. There has been preserved in the small Shōdō Museum in Tokyo an important ceramic 'document' in the form of a spouted bowl, bearing an incised inscription round the rim which gives the name of its maker and the date: the third year of Chung-p'ing in the Han dynasty, corresponding to A.D. 186. This is accepted as authentic by Japanese authorities and, though it appears at first to be unglazed, close inspection shows that it is speckled with either glaze or natural kiln gloss. Koyama maintains that this is the remains of a primitive ash-glaze which, owing to insufficient firing, has not properly fused and is 'clotted on the surface like powder'. The inscription appears to be genuine, but the Yüeh attribution seems only tenable on the assumption that a celadon glaze would have developed under appropriate firing conditions.[1]

It should be explained at this point just what is meant by Yüeh ware. Among the early porcelains mentioned in the T'ao-shuo and other literary sources, none achieved greater fame than 'the porcelains of Yüeh-chou'. Dr. Ch'ên lists ten poets writing from about 757 to 900 whose verses make mention of Yüeh ware,[2] and a celebrated couplet by the ninth-century poet, Lu Kuei-mêng, runs as follows:

[1] Fujio Koyama, 'The Yüeh-chou Yao Celadon excavated in Japan', *Artibus Asiae*, Vol. XIV, 1-2, 1951, Fig. 2.
[2] Wan-li Ch'ên, *Chung-kou Ch'ing-tz'u Shih-lüeh* (Outline History of Chinese Celadon), Shanghai, 1956, p. 8.

B. Incense burner on three legs. Chinese. Lung-ch'üan ware (Kinuta type). Sung dynasty. Height $4\frac{1}{8}$ in. Diameter $5\frac{1}{8}$ in. British Museum.

'Among the autumn winds and dew, Yüeh-yao is born,
Lifted from a myriad peaks, the blue-green colour comes. . .'

As Brankston put it when Yüeh ware had at last been identified:
'Now we may see this colour of distant mountains and feel the
winds and dew, for Yüeh ware has been dug up and is back among
the living wares again'.[1] Yüeh-chou is the ancient name for Shao-
hsing and, in the opinion of J. M. Plumer, may well have referred
to the general area, even as far as the lake, Shang-lin Hu, on which
we shall see that the main centre for the production of Yüeh ware
was established.[2] However, by general consent the term Yüeh has
nowadays been extended still further to cover most of the early
celadon wares manufactured in Chekiang Province, and this has
involved its being pushed back in time to the Six Dynasties and
even to the Later Han.

An enormous quantity of Six Dynasties Yüeh ware has been
recovered from numerous graves near Shao-hsing and elsewhere
in Chekiang and, although the excavations were mostly illicit, the
finds also included bricks and bronze mirrors stamped with year-
marks of the Six Dynasties period, so that their approximate date
can hardly be questioned. Koyama states that some two or three
thousand graves were opened and ransacked, with the result that
the antique shops in Hang-chou and Shanghai were crowded with
the wares. They consisted of jars, dishes, bowls, basins, wine-cups,
ladles, light-stands and various kinds of mortuary vessels. Many
of these were acquired by Japanese collectors in the late thirties.

The most celebrated of these pottery wares found at Shao-
hsing is a green-glazed vase standing about 40 cm. high with an
elaborate group of figures modelled in the round and applied
above the shoulders, encircling the long neck: these comprise
buildings with Chinese roofs, sometimes in several tiers, musicians,
flying birds, dogs—while all round the shoulders are attached
figures of animals, tortoises and fish. One of these is a tortoise

[1] A. D. Brankston, op. cit.
[2] J. M. Plumer, 'Certain Celadon Potsherds from Samarra traced to their
Source', *Ars Islamica*, Vol. IV, 1937, p. 195.

bearing a tablet with the inscription: 'In the third year of Yung-an (corresponding to A.D. 260) you are wealthy and prosperous, qualified to be a minister and a procreator enjoying longevity and endowed with talents which have no equal in history'. This remarkable piece is now installed in the Palace Museum at Peking and is regarded as an early specimen of Yüeh ware. The glaze is described by Dr. Ch'ên as 'deep in colour and thick in consistency, marking a great improvement in firing technique and showing that celadon ware had reached an important stage in Chinese ceramic history.'[1]

Another early dated specimen mentioned by Dr. Ch'ên is a lion-shaped vessel found in a tomb just outside Nanking in 1954: this bore the date '14th year of Ch'ih-wu', corresponding to A.D. 251, inscribed on the side. A lamp in the form of a bear also came from a tomb at Nanking and bore a date corresponding to 265; a large quantity of celadon was found in this tomb, including a remarkable vessel in the form of a ram with finely incised detail. At I-hsing in Kiangsu Province a tomb was found to contain forty-two pieces of celadon, including jars, basins, wine-flasks, winged wine-cups and two globular incense burners with triangular perforations in the side surmounted by a knob on which a small bird was perched;[2] many of the bricks used in constructing this tomb were dated '7th year of Yüan-k'ang', or 297. Other graves in the same province dated 273 and 366 likewise yielded celadon wares. In Chekiang Province a large number of graves have been excavated in recent years around Huang-yen and found to contain large quantities of celadon ware—Hsien-ming Fêng mentions fifty-two excavated in 1956, many of which were dated from A.D. 276 to 462.[3] Numerous other graves have been

[1] Wan-li Ch'ên, *op. cit.*, p. 4.
[2] See William Willetts, *op. cit.*, pl. 154; the date is incorrectly shown as A.D. 279.
[3] Hsien-ming Fêng, 'Important Finds of Ancient Chinese Ceramics since 1949' (in Chinese), *Wen Wu*, No. 9, pp. 26–56, illus., p. 28. See translation by Mrs. Hin-cheung Lovell, produced by the Oriental Ceramic Society and the Victoria and Albert Museum, Chinese Translations, No. 1, 1967.

found in Fukien, Kiangsi, Hunan and Szechwan Provinces containing celadon wares and datable to the fourth and fifth centuries. Dr. Ch'ên refers to celadon mortuary wares excavated from graves with the following Six Dynasties epitaphs:

T'ai-k'ang	(Western Chin—A.D.			280–290)
Yüan-k'ang	(,,	,,	,,	291–301)
Yung-k'ang	(,,	,,	,,	300–302)
Chien-hsing	(,,	,,	,,	313–318)
T'ai-hsing	(Eastern Chin—		,,	318–322)
Hsien-ho	(,,	,,	,,	326–335)
Hsien-k'ang	(,,	,,	,,	335–343)
Chien-yüan	(,,	,,	,,	343–345)
Yung-ho	(,,	,,	,,	345–357)
Shêng-p'ing	(,,	,,	,,	357–362)
T'ai-ho	(,,	,,	,,	366–372)
Ning-k'ang	(,,	,,	,,	373–376)
T'ai-yüan	(,,	,,	,,	376–397)
Yüan-chia	(Liu Sung		,,	424–454)
T'ien-chien	(Liang		,,	502–520)
Ta-t'ung	(,,		,,	527–550)

Besides food vessels and model buildings, including sheep-pens and pigsties, the mortuary wares comprised human, bird and animal figures; pots and jars with two, four or six 'ears'; wine-pots modelled after bronzes, with animal masks and rings at the sides; basins and lamp-stands; platters with bear-legs and 'chicken-head' ewers. One of the finest and most unusual specimens is a model of a man astride a *ch'i-lin* and wearing a tall, cylindrical hat.[1] Discoveries of this kind continue year by year, and Dr. Ch'ên aptly comments: 'The large-scale production of green-glazed

[1] See *The Selected Porcelains from the collection of the Palace Museum*, Wen Wu Press, Peking, 1962.

wares in the period of the Southern Dynasties[1] may be said to have laid the foundations for rapid development in the ensuing Sui and T'ang periods'.[2]

The first discovery of a Yüeh kiln centre dating from the Han period was made by Mr. Tsuneo Yonaiyama, Japanese consul at Hang-chou, in 1930. This was at Tê-ch'ing, some twenty-five miles north of Hang-chou. Mr. Yonaiyama had been intrigued by the name of a village, Ho-yao, because the word *yao* signifies 'ware' or 'kiln', and visited the locality by boat along one of the numerous canals. After making inquiries, he found an old man who 'knew of a place where bowls were once made' and was led by him to the lower slopes of a mountain, where the rice-fields ended amidst pine-woods. The upturned soil of the fields was littered with fragments of early celadon wares, especially bowls and chicken-ewers (Plate 1). Most of these were covered with a grey-green celadon glaze, but some others had a black *temmoku* glaze—the earliest examples of their kind found at a kiln site.[3] From the style of make and the quality of the glaze there is little doubt that the kilns originated in the Han period and were active throughout the Six Dynasties.

Mention has already been made of Brankston's visit to another important kiln centre at Chiu-yen, near Shao-hsing. This site had been first inspected a year earlier by Mr. Yuzō Matsumura, successor to Yonaiyama as Japanese consul at Hang-chou, since excavated articles were already appearing on the Hang-chou market. Matsumura likewise made his way to the site by boat and discovered a large heap of fragments and saggars in a bamboo-grove, where there had evidently been one or more kilns. Most of the pieces were from bowls, jars and pots, but he also collected a figurine and two ink-pallets. The majority were decorated with

[1] Northern and Southern Dynasties is another term used for the Six Dynasties.

[2] Wan-li Ch'ên, *op. cit.*, p. 5.

[3] A black chicken-ewer found at Tê-ch'ing is identical to one recovered from a tomb at Hang-chou dated 364; and a celadon lid is exactly like two from tombs at Nanking dated 345 and 348 (Hsien-ming Fêng, *op. cit.*, p. 27).

diamond diaper patterns, often encircling the exterior just below the rim in a band about one inch wide. Some of the pieces had lions' or chickens' heads in relief on the shoulders. The glaze was olive-green for the most part, or the 'putty colour' which characterizes many early examples of Yüeh ware. It seems that Brankston did not actually see the site in the bamboo-grove in the following summer, as he mentions that most of the pieces he collected were obtained from villagers or 'fished up' with a hoe from the river.

The wares made at the Chiu-yen kilns are very distinctive. Many of the bowls and jars have concave bases with brown 'haloes' where they rested on irregular lumps of clay during the firing, and there was a strong tendency to grotesquerie: besides the jars and ewers with their animal-masks or chicken-head spouts there are many fantastic animal forms, such as light-stands modelled as rams, lions or bears (Plate 4b). A good selection of these are described and illustrated in a study by Karlbeck of the Ingram Collection, now at the Ashmolean Museum in Oxford, which is probably the largest in existence outside China,[1] and it is likely that many of the mortuary vessels recovered from graves near Shao-hsing came from the Chiu-yen or other nearby kilns. In recent years Chinese archaeologists have discovered additional kilns in the same region at Wang-chia-lü, south of Chiu-yen, and at Shang-tung, near Hsiao-shan, about twenty miles to the west, where celadons very similar in form, glaze and style were manufactured. The attribution of surviving specimens to specific kilns is always risky unless they are actually wasters recovered from the site, and criticism has been expressed of the Japanese tendency to identify pieces with no history as the products of particular kilns. However we must recognize that the Western habit of using the names of wares descriptively, regardless of their provenance, is not normally favoured in China and Japan, and the usual method in these countries is to assign pieces to actual kilns wherever the body,

[1] Orvar Karlbeck, 'Proto-porcelain and Yüeh Ware', *Transactions of the Oriental Ceramic Society*, Vol. 25, 1949–50.

glaze and general style seem to warrant such attributions. Difficulty obviously arises in a case like this, where similar wares were made at several groups of kilns in the same area, but until further study has enabled distinctive features to be isolated, it seems permissible to classify the general type as Chiu-yen ware from the name of the earliest and perhaps the most important kiln centre in the district.

Such evidence as is available dates the Chiu-yen and neighbouring kilns to the Six Dynasties era, but there is good reason to believe that the period of activity extended on into the T'ang period. Koyama's provisional classification of Tê-ch'ing as a Han/Six Dynasties centre and Chiu-yen as Six Dynasties/T'ang may be accepted, anyway for the time being. There is one type of ware common to both sites—doubtless since found also at others—which should be specially mentioned, though it was evidently made only on a limited scale (Plates 3 & 4a). In this the celadon glaze is spotted or splashed with brown—an early form of what the Japanese have called *tobi seiji*, or 'spotted celadon'. This was produced by brushing iron oxide onto the surface before the firing, resulting in a style of ornament which is very striking and popular in the Far East.

The largest and most important Yüeh kiln centre was discovered near Yü-yao by another Japanese, Mantarō Kaida, working on the instructions of a perspicacious scholar, Dr. Manzō Nakao, early in 1930. Dr. Nakao had found a clue to their location in the local annals of the Chia-ching period (1522–66), which were quoted in the district guide, and his suspicions were justified when Kaida returned from his expedition with a large collection of shards and wasters. These had been found on the shores of a large lake, Shang-lin Hu, where a great number of kilns had been active throughout the T'ang period. This discovery was of the greatest significance, since it resulted in the identification of Yüeh-chou ware, hitherto known only from Chinese literary sources. Dr. Nakao sent some of the shards to the British Museum, enabling Hobson to identify several complete specimens in Eng-

lish collections and throwing a flood of light on the early development of celadon ware.

Numerous scholars followed in Kaida's footsteps, including
Yonaiyama, Matsumura, Plumer, Brankston, Dr. Ch'ên and
Dr. Nakao himself. The scene was vividly described by Yonaiyama as follows: 'What a lovely lake this is! The mountain keeps
its age-old calm; pink azaleas are in full blossom on its slopes and
pine-trees stand here and there among the flowers. The exposed
surface of the rock shines white. The lake is long and narrow, and
reeds grow along its margin. On the shores there are kilns for
making bricks. Clay is dug from the shallower parts of the lake
and carried to the kilns by boat. It is a quiet spot, surrounded by
mountains. On all sides are white rocks, green pine-trees, pink
azaleas and the blue-green ranges. . . .'[1]

The pottery fragments were scattered among the cultivated
fields and foothills of the mountain at the southern end of the lake
and were in such numbers as to form small hills. Matsumura
likened the scene to 'an ancient battlefield', with countless fragments of bowls, jars, ewers, many fused together by the heat of the
kilns. In addition there were masses of kiln waste, stilts, saggars
and other debris. Fragments of celadon wares had been built into
the walls of a small farm-house beside the lake. Matsumura states
that he found two grave-tablets inscribed with Han period dates,
but it is unlikely that the kilns were constructed as early as this;[2]
however, the discovery nearby of a celadon tablet bearing an
epitaph dated the third year of Ch'ang-ch'ing, or 823, proved that
the kilns were in operation during the T'ang period.[3] The finest
wares are believed to have been made in the Five Dynasties period
(907–60) and on into the early part of Sung (Plate 5a).

A characteristic feature of Shang-lin Hu wares is the high,

[1] Tsuneo Yonaiyama, *Shina Fudoki* (Chinese Record), Tokyo, 1939, p. 397.
[2] Some newly-discovered sites at Ao-ch'un-shan nearby are thought to date
from the fourth century (Hsien-ming Fêng, *op. cit.*, p. 26).
[3] Fujio Koyama, 'A study of a T'ang Epitaph. . . .'(in Japanese), *Tōji*, Vol.
VIII, No. 5, October 1936; an English translation of this article appeared in the
Far Eastern Ceramic Bulletin, No. 12, December, 1950.

splayed foot, but the decoration, sometimes moulded but more often incised with a fine point, is the most distinctive evidence that the manufacture of celadon ware had reached a new peak (Plate 5b). As Dr. Ch'ên puts it: 'The rich variety of designs . . . is unprecedented in the history of Chinese porcelain', consisting of 'butterflies flying in pairs, parrots and phoenixes, chirping birds, among blossoms and storks soaring high in the cloud'.[1] The decorative designs of the T'ang period found full expression in these lovely wares, and the glaze too had reached a new high point, being of a quality and lustre far surpassing that of Six Dynasties Yüeh ware.

EARLY YÜEH KILN SITES

Further investigations have been carried out in recent years by Chinese archaeologists, not only at Shang-lin Hu but at the neighbouring lakes, Shang-ao Hu and Pai-yang Hu, as well as on the slopes of nearby hills. The deposits were up to six feet in depth, and over one thousand shards and wasters were collected, several of which bore inscriptions with the dates 850, 922 and 978.[2] About twenty separate sites have been found altogether at or near Shang-

[1] Wan-li Ch'ên, *Yüeh Ch'i T'u-lu* (Illustrations of Yüeh Ware), Shanghai, 1937, p. 3 of English preface.
[2] Tsu-ming Chin, 'The Celadon Kiln Sites of Yü-yao, Chekiang' (in Chinese with English summary), *K'ao-ku Hsüeh-pao*, No. 3, 1959.

lin Hu, according to Fêng, the best pieces coming from Kou-t'ou-ching-shan, Ta-pu-t'ou and Ch'ên-tzŭ-shan, while the largest quantity of remains was at Huang-shan-shan and Yen-tzŭ-k'un. About twenty miles to the west another large group of kiln sites was found near Shang-yü—Fêng mentions six localities which made celadon wares in the Six Dynasties and five during the T'ang/Five Dynasties. There were fragments of ewers, bowls, basins and ink-stones, many of them decorated with a wide variety of impressed and incised motifs such as waves, lotus petals, dragons, phoenixes and birds; while a few miles south-west of Huang-yen, near the sea-coast and about one hundred miles further south, kiln sites have been discovered at eight places. These Huang-yen kilns made high-quality wares after the same style as Yü-yao, or Shang-lin Hu, and the fragments, which were mainly of covered boxes and bowls or dishes, are described as thinly potted and having a very hard porcellanous body with a beautiful 'moss-green' glaze. Many of them bore exquisite decoration incised with a fine point of such motifs as phoenixes, parrots, Mandarin ducks, peonies, chrysanthemums, lilies and lotuses. They are among the finest Yüeh shards ever recovered. Another kiln centre at Yü-wang-miao, close to Shao-hsing, is reported to have made high-quality wares with finely incised ornament, but this at present lacks substantiation, though the site is among those listed by Dr. Ch'ên. Numerous other Yüeh kiln sites have been found in Chekiang Province and are the subject of brief reports listed by Fêng, such as Lü-pu-k'êng, near Li-shui; Hsia-pi-shan and T'an-t'ou, near Yung-chia; Hsiao-pai-shih, near Yin (or Ning-po); Hsia-p'u-hsi, near Shao-hsing; and Hsi-shan, near Wên-chou. In addition, similar wares were made in Kiangsu Province at Chün-shan, near I-hsing, and far to the west in Szechwan—at Yü-huang-kuan, near Hsin-chin; at Ku-i and two other localities near Ch'iung-lai (both Hsin-chin and Ch'iung-lai are about 50/60 miles west of Ch'êng-tu); and at Liu-li-ch'ang, near Hua-yang, just south of Ch'êng-tu.

By the T'ang period Yüeh celadon ware had improved beyond

all measure; the shapes were easy and flowing, while the glaze was smooth and even; but it was not until the Five Dynasties period that the high-water mark was reached. T'ang poets united to praise the beauty of the designs and glaze, which was known as *pi sê*, or 'secret colour', and highly prized for its resemblance to jade and the mysterious quality with which it invested tea-bowls and other wares; and its fame grew until it was reserved for private use by the princes of the Wu-Yüeh ruling house. There is a theory that the legendary Ch'ai ware, about which nothing definite is known, was in fact nothing more than superior Yüeh ware made for presentation to the Later Chou, last of the Five Dynasties, which became known by the family name of its ruler Shih Tsung, son of Ch'ai Shou-li.[1]

Dr. Ch'ên states: 'The general populace was prohibited from using these porcelains. Before despatching the tribute to his over-lord, Ch'ien Shu arranged the pieces one by one in his garden and burned incense before them to testify his fidelity to his feudal lord. He endeavoured to maintain his position by securing the goodwill of his overlord by such acts as this. It was for this reason that the general use of Yüeh ware was proscribed; for these acts demonstrated the preciousness of the ware, which was for the exclusive use of his lord'.[2]

The princes of Wu-Yüeh ruled at Hang-chou from 907 until 978, after which the state was absorbed by the growing power of Sung, and fifty thousand porcelains, one hundred and fifty having gold rims, were made for the Emperor in this year in token of submission. Dr. Ch'ên also lists twelve references in Chinese literary records to the presentation of Yüeh wares, many of them ornamented with gold, as tributary gifts between the years 924 and 983.[3] After 978 it seems that government supervisors were appointed, but the loss of patronage by the local rulers resulted in a steady decline, though wares continued to be made for

[1] Wan-li Ch'ên, *op. cit.*, (*Chung-kuo Ch'ing-tz'u Shih-lüeh*) p. 19.
[2] *Ibid.*, p. 16.
[3] *Ibid.*, p. 15.

imperial use at least until the year 1068.[1] The potters gradually drifted away to new and thriving potteries around Lung-ch'üan and the new Sung capital at K'ai-fêng in the north. Some of them are thought to have migrated and settled in Korea, where the earliest celadon wares were now being made, since it is evident that these were under strong Yüeh influence and relied on techniques introduced from Chekiang, which was within easy range of the south-west coast of the peninsula. Indeed, the first great Korean celadon pottery centre was established just south of Kangjin, the nearest point to the Yüeh factories.

The scene thus shifts from South China to North China and Korea, though it was not long before the new potteries established around Lung-ch'üan, also in Chekiang, achieved a success which eclipsed even that of Yüeh ware and the celadons made in the north. Indeed, some wares which have been regarded as Yüeh celadons of the early Sung period are believed by Dr. Ch'ên and other Chinese authorities to be an early form of Lung-ch'üan ware; but this only serves to emphasize the indebtedness of the Sung period celadons to the first achievement in this class of ware, which undoubtedly belongs to the Yüeh potters.

Before ending this brief historical survey of Yüeh ware, it is necessary to say a few words about its role as the first great Chinese export ware; for in the T'ang period Chinese junks plied as far as the Persian Gulf and Arab writers from the ninth century make reference to the large-scale importation of Chinese porcelains. Yüeh fragments have been found at Brahminabad in India, at Samarra on the River Tigris (abandoned in 883) and at Fostât, the old city of Cairo, which was at its zenith in the ninth century, and complete specimens have been preserved from this period in Japan and other parts of the Far East. It seems that a few examples of the ware reached neighbouring countries during the eighth century, but the period when it became a staple export probably began in the ninth century and continued throughout the T'ang dynasty and onwards into the Sung.

[1] Wan-li Ch'ên, *op. cit.*, p. 20.

The success of any new type of pottery is naturally followed by its wide-scale copying, not only in the country of origin but also elsewhere, and this was very marked during the interval between the collapse of the T'ang dynasty and the rise of Sung. We find a great variety of celadon wares developing at potteries all over China during this period, and the exact provenance of many of these has still to be determined. The late Sir Herbert Ingram was deeply interested in this line of study, though he could only pursue it on the basis of specimens he had collected or examined in other collections; and more fruitful work is now being conducted in China itself as a result of archaeological excavations. Unfortunately most of the Chinese reports have been very terse, while the illustrations are often so poorly reproduced that it is difficult to reach any positive conclusions; and in many cases we are informed that a kiln site has been discovered at such-and-such a place but no report has yet been published. However, despite the lack of data, it seems worth listing some of the types of celadon ware which have been provisionally classified into groups, either from studies of specimens collected or on the basis of field-work in particular areas—including a few which are known only from literary references and have not yet been definitely identified. Most of these still problematical wares date from the late T'ang period onwards into the early Sung:

(1) *The 'Grey Ware'*, so-named by Sir Herbert Ingram, which is so close to the known types of Yüeh that it was probably made at one of the Yüeh kilns, though it has not been possible to identify this from the Chinese archaeological reports. The forms are most elegant and are enhanced by a very pale bluish or greyish green glaze, which appears almost white on the light-coloured body. The possibility exists, as mentioned above, that it is an early form of Lung-ch'üan ware.[1]

(2) *Wu-chou Ware*, mentioned in the 'Tea Classic' of the T'ang writer, Lu Yü. Dr. Ch'ên states that a kiln site was discovered

[1] See the author's book: *Chinese Celadon Wares*, London, 1958, p. 19.

to the west of Chin-hua (the modern name for Wu-chou) in Chekiang with fragments of celadon ware which were rather thickly potted and poor in finish but dated from the T'ang period.[1] Fêng mentions a nearby site, Wu-chu-t'ang, at which celadon shards mottled with brown were found but were datable to a much earlier period, about the fourth century, though the kilns apparently continued to the T'ang and Five Dynasties. It has not yet been established whether these made the Wu-chou celadons of literary fame. No detailed reports have yet been published.

(3) *Yo-chou Ware*, also mentioned in the 'Tea Classic' and the subject of detailed studies by Dr. Isaac Newton on the basis of a collection of Hunan ceramic wares made by him in Hongkong during the years 1947–50.[2] There seem to have been three main kiln centres in Hunan Province making a wide variety of wares with brown, yellow and green glazes: around Yo-yang; at Hsiang-yin, some fifty miles further south; and also at Ch'ang-sha, the provincial capital. Some of the celadons bear a close remblance to Yüeh ware, of which they are no doubt country cousins. Fêng describes the wares made at Wa-cha-p'ing, near Ch'ang-sha, as 'unparalleled for inventiveness in glaze colour and decoration' and enumerates eleven varieties, including celadons with brown and green mottling, painted and applique decoration, etc., all dating from the T'ang/Five Dynasties. These Hunan wares are usually rather brittle and the glazes have often suffered damage by flaking.

(4) *Shou-chou Ware*, also mentioned in the 'Tea Classic', seems to fit the description of wares with a yellow glaze found at kiln sites near Huai-nan Shih in Anhui Province—at Ma-

[1] Wan-li Ch'ên, *op. cit.*, p. 13.

[2] Isaac Newton, 'Some Ceramic Wares reportedly excavated near Changsha', *Far Eastern Ceramic Bulletin*, Vol. V, 1953; also 'Chinese Ceramic Wares from Hunan', *Far Eastern Ceramic Bulletin*, Vol. X, 1958; also 'A Thousand Years of Potting in the Hunan Province', *Transactions of the Oriental Ceramic Society*, 1950–51, Vol. 26.

chia-kang, Shang-yao Chên, Yü-chia-kou and Wai-yao—
described by Fêng as 'nearer the northern than the southern
style of ceramic wares'.

(5) *Hung-chou Ware*, likewise mentioned in the 'Tea Classic'.
The attribution of bowls or dishes with carved lotus decoration at the centre, inside, under a light olive-green glaze
found in some Japanese and Western collections never
rested on a secure basis and is now discredited by Chinese
scholars (see Fêng, *op. cit.*, p. 29). Several of these specimens
were acquired in Nan-ch'ang (the modern name for Hung-
chou) in Kiangsi Province and a considerable number have
been found in tombs in the vicinity dating from the Six
Dynasties period, but no kiln sites have yet been located.

(6) *Ching-tê-chên:* an olive-coloured ware was found in some
quantity at Shih-hu-wan, some miles from Ching-tê-chên,
in Kiangsi Province, also on the slopes of a hill, Yang-mei-
t'ing, further to the south. Dr. Ch'ên states that it has a thin
glaze and very fine decoration, while the shapes and style are
characteristic of the T'ang period.[1]

(7) *Chi-an:* the site of the famous Chi-chou ware of the Sung
dynasty was found at Yung-ho Chên, near Chi-an in Kiangsi
Province, and it seems that celadon was a secondary product
of these kilns. Dr. Ch'ên states that he found 'a fairly large
number of celadon fragments' in the shallow waters of the
Kan River nearby. The glaze was dark green and there were
indications that the kilns dated from the T'ang period.[2]

(8) *Li-shui Ware*: Chinese literary sources mention this ware,
made at Li-shui, between Lung-ch'üan and Ch'u-chou in
Chekiang Province, in early Sung times. Some pieces were
said to be dark and others light. Koyama believes that the
glaze was often olive-coloured and that relief decoration was
carved underneath. This seems to fit some indeterminate
celadons of early date very well, and the writer has suggested

[1] Wan-li Ch'ên, *op. cit.*, p. 13.
[2] *Ibid.*, p. 14.

that a tentative attribution to Li-shui would be more appropriate than the former arbitrary classification as Yüeh or Northern Celadon according to personal preference.[1] Sir Herbert Ingram drew attention to one feature many of these pieces have in common: the glaze has been roughly scraped off the base before the firing. Fêng states that celadon wares were produced at Lü-pu-k'êng in the vicinity from the Six Dynasties period on into the T'ang, but no complete report has been published.

(9) *The 'Pale Celadon Ware'*, so-named by Sir Herbert Ingram, who owned six or seven pieces and had noted several others in Western collections. The chief characteristics are a pale grey body, 'a light-coloured celadon glaze quite different to that of Lung-ch'üan ware' and 'a double-bevelled foot-rim'. While the date of make is clearly early Sung, the provenance is unknown, though it is likely that similar specimens have been found at sites investigated by Chinese scholars in recent years.

(10) *'Precursory Ju Ware'*: several bowls owned by Sir Herbert Ingram with lobed sides and a close-fitting grey-green glaze have been suggested as possible prototypes of Ju ware. If this theory is correct, they were no doubt made at Lin-ju in Honan Province (the modern name for Ju-chou).

In addition to the above, some mention should perhaps be made of the rather ordinary class of celadon ware often seen in Japan and there known as *Shuko seiji*. It has an olive or yellowish tone and incised decoration inside, also comb pattern—often called 'cat-scratch' in Japan—on the exterior. Yonaiyama found shards of this ware at Tê-ch'ing, mostly fragments of bowls and dishes, and it was probably made at a number of potteries in Chekiang and also perhaps in Fukien, where the shards recovered in recent investigations seem to be of related types. The 'loquat-coloured' glaze is highly esteemed in Japan, where there are many examples

[1] See *Chinese Celadon Wares, op. cit.*, pp. 23-4.

believed to date from the Southern Sung period which were imported in the thirteenth and fourteenth centuries.[1]

Although most of the important early celadon kiln sites have been mentioned above, the list has by no means been exhausted and indeed lengthens with new discoveries every year. While Chekiang was undoubtedly the main centre, large new fields of study have been opened up by finds in other provinces such as Kiangsi, Hunan and Szechwan; and at least a dozen kiln sites each in Fukien and Kwangtung have been located where celadon wares were made either primarily or as a secondary product among other types of ware. Some of these kilns were started in the T'ang period and most of them were active in the Sung. They are listed by Fêng, but reports for the most part have been scanty.

[1] Fujio Koyama, *Shina Seiji Shi-kō* (Sketch History of Chinese Celadon), Tôkyô, 1943, pp. 309–10.

CHAPTER THREE

Chinese Celadon Wares in the North and in the South

It was natural that the achievements of the Chekiang potters should lead to efforts being made to produce celadon ware in the north; and excavation of a tomb at An-yang in Honan as long ago as 1929 proved that at least one type of celadon was in production in North China in the Sui period (581–618). The tomb was found to contain an epitaph giving the name of the man buried there, which was Pu Jên, and the date of his death, 603. In addition, it contained various mortuary articles including ten celadon wares; these comprised four jars, each having four handles on the shoulders and small lids, five small bowls and a high-footed stand, all made of buff stoneware covered with a brownish green celadon glaze. The jars were of distinctive shape, swelling at the waist, and only the top half was glazed. Koyama remarks that he often noticed this kind of Sui celadon in antique shops in North China around 1942, and there are quite a number of examples in Japanese and Western collections (Plate 7). Quoting Koyama further: 'These Sui celadons resemble the Han and Six Dynasties Yüeh ware in their robust simplicity, their strange and archaic appearance. They are thickly potted and roughly made, or in a word of poorer quality than Yüeh ware. I have never seen Sui wares with any decoration or displaying refined technique: they all seem to have been made without much care.'[1] There is very little information as yet regarding the kiln sites where these wares were

[1] Fujio Koyama, *op. cit.*, p. 42.

4

made, but a preliminary study of a Sui kiln site at Chia-pi-ts'un in the Tz'ŭ-chou area has disclosed one possible source.[1]

Dr. Ch'ên is unable to answer the question whether there was any connection between these early Sui celadons and the accomplished ware produced in the Sung period, probably at several potteries in Honan and Shensi, which has come to be known in the West under the generic term of Northern Celadon. There is, indeed, a sound basis for this description, vague as it undoubtedly is, since the ware is associated with the traditional Tung celadon of the north, said to have been made at Chên-liu near K'ai-fêng—where Northern Celadon fragments have been found—and there is no question that it originated in North China. However, excavations carried out in recent years near T'ung-ch'üan, about fifty miles north of Sian in Shensi Province, have revealed one of the main centres for production of a large variety of Northern Celadon and Tung-type wares with an olive or yellowish tone.[2] It is possible, even from illustrations, to identify as likely products of these Yao-chou kilns—the traditional name from the former township or district in which they were located—some of the better-known examples of Northern Celadon ware in Western collections; but until further discoveries are made at proven kiln sites in North China it will be best to maintain caution and to regard the whole problem of Northern Celadon as one needing a great deal of clarification.

Another good reason for conservatism is the confusion which persists in regard to nomenclature between Chinese, Japanese and Western authorities. Many Northern Celadons are known to have been made at Lin-ju hsien, the modern name for Ju-chou, and the mainland Chinese, supported by the Japanese, call these either Lin-ju or Ju ware, regarding them as 'the ordinary type made for common use', whereas an entirely different tradition has been

[1] Hsien-ming Feng, 'Preliminary Study of a Sui Celadon Kiln Site at Chia-pi-ts'un, Tz'u-hsien, Hopei' (in Chinese), K'ao-ku, No. 10, 1959.

[2] See Yao-tz'u T'u-lu (Illustrations of Yao Porcelain), Shensi Provincial Museum, Sian; Peking, 1956; also Excavations of the Yao-chou Kiln Sites at T'ung-ch'üan, Shensi, Peking, 1965.

maintained by the custodians of the Palace Collection in Taiwan and seems to be confirmed by the literary evidence, which is also considered decisive in the West. According to this, there is only a single category of Ju ware, the superior class made for the imperial court on a very small scale and known by all authorities, East and West, as Imperial Ju. The Northern Celadons found at Lin-ju hsien are entirely different in style, having carved decoration under a glaze of decidedly olive tone, whereas Imperial Ju, with rare exceptions, is devoid of decoration, relying solely on purity of form and subtlety of glaze, which together confer a beauty and distinction almost without peer in the ceramic products of any land and age. To make matters still more complex, Dr. Ch'ên of Peking states that plain, 'onion-green' shards with 'broken-ice crackle' are found north-east of Lin-ju hsien and that these represent the early type from which Imperial Ju was developed.[1]

Before the advent of Ju, the celebrated white porcelains of Ting-chou were used in the palace. However, according to Chinese literary sources,[2] certain imperfections gave rise to dissatisfaction—Dr. Ch'ên suggests that the unglazed rim caused by their being fired upside-down may have been chief among these—with the result that the new and superior celadon (ch'ing) wares made at the Ju-chou kilns were employed in their place. Another early source[3] states that the Kuan, or official, factory was established at the capital, now known as K'ai-fêng, in the Chêng-ho era (1111–17). It is of no small significance that celadon ware

---

[1] Wan-li Ch'ên, op. cit., p. 36.

[2] The earliest of these is the *Lao-hsüeh An Pi-chi* by Lu Yu (1125–1210), 1877 ed., Ch. 2, f. 6b; the passage appears also in the fourteenth century *Shuo Fu* by T'ao Tsung-i, 1930 ed., Ch. 4, f. 22b. It seems that this passage was the basis for a slightly different version which appeared in the *Fü-hsüan Tsa-lu* by Ku Wên-chien, published between 1260 and 1279, as also quoted in the *Shuo Fu*, 1930 ed., Ch. 18, f. 10b. This reference was erroneously attributed to the *T'an Chai Pi-hêng*, but the mistake is explained by Sir Percival David in his 'A Commentary on Ju Ware', *Transactions of the Oriental Ceramic Society*, Vol. 14, 1936–7, pp. 27–9.

[3] The *Fü-hsüan Tsa-lu, op. cit.*, as quoted in the *Shuo Fu*, 1930 ed., Ch. 18, f. 10b.

had thus risen in favour to the point where it supplanted the time-honoured white porcelains of Ting-chou and became the official ware used in court and palace.

The exact date when Ju became the official ware is not known,[1] nor is the period of its supremacy. However, it could not have out-lasted the destruction of the Sung capital by the Chin Tartars in January, 1127. Indeed, there are grounds for holding that 'Northern Kuan'—so called to distinguish it from the later Kuan ware made in the south—shortly underwent a further change, either to a variety of wares made by skilled potters from different kilns (limited by some authorities to wares of Ju type) or to a special class of ware made at the capital, possibly Tung.[2] In any case the important consideration from our point of view is that all these wares were celadons, and the later products of Hang-chou, which are known as Southern Kuan and became the official ware of the Southern Sung court likewise were celadons. In fact, it was not until the rise of blue-and-white and polychrome wares in the Ming period (1368–1644) that any other porcelains were selected as imperial wares.

A few words of description and explanation will not be out of place before leaving the subject of Ju ware, since its unique beauty and rarity have captured the imagination of ceramic students all over the world. Indeed, considering the fact that specimens were said to be 'very hard to procure' as long ago as 1192, while at the end of the sixteenth century a Governor of Ju-chou was able to see only a single piece, it is rather surprising that at least sixty examples are known to have survived—thirty-three in Western collections, twenty-three in the Palace Collection in Taiwan,[3] one piece in Japan and a few in Peking. Apart from those in Taiwan, the

[1] The year 1107 is given by most modern writers but rests on the sole evidence of an inscribed ritual disc in the Percival David Foundation, the authenticity of which is open to question.

[2] For a detailed discussion of these difficult questions see *Chinese Celadon Wares*, op. cit., Chapter 4: 'Ju and Other Northern Wares'.

[3] As shown in *Ju Ware of The Sung Dynasty* (see Bibliography); there may also be others.

largest number—fourteen pieces together with two miniatures and eight fragments—are in the Percival David Foundation in London and should be examined by all who take any interest in oriental pottery, or indeed in ceramics generally.

Ju ware is distinguished above all by its fine potting and lovely celadon glaze (Plate 8). Opinions will differ concerning the precise range of colour, which may be broadly described as a smooth, opaque bluish green, often considered to have a tinge of lavender. Chinese experts have classified the shades into three main groups, 'sky-blue', 'pale blue' and 'egg-blue'—but these all employ the term *ch'ing*, which signifies either blue or green, and the fact is that, like many other celadon wares, Ju will appear to have a different tone according to the conditions of lighting and, perhaps, the frame of mind of the viewer. This variable colouring is not the least of its charms. Another well-known characteristic is the presence of fine crazing, which some have taken to be the first deliberate crackle produced in China. However, it seems more likely that the Chinese authorities in Taiwan are correct in pronouncing it fortuitous, occurring 'when the thickness of the glaze was not constant or as a result of either over-firing or under-firing during manufacture: a piece which was technically perfect and was fired correctly should have no crackle'.[1] The *Ko-ku Yao-lun*, published in 1387, remarks that 'genuine as are the ones with crackle, those without it are still better'. A flaky texture is a frequent though not invariable feature; and the presence of neat, small spur-marks on the base where the vessels were supported in the kilns is distinctive (Plate 9). The body was made of fine clay said to contain traces of copper oxide and has acquired a pale buff tinge in the firing. It is described by Sir Harry Garner as more like pottery than porcelain, perhaps as a result of being fired at a lower temperature than was customary for Sung porcelain.

At least twenty years before Ju ware had been identified, anyway in the West, Hobson wrote that it 'was evidently of the grey

[1] *Ju Ware of the Sung Dynasty*, (see Bibliography), p. 22.

green celadon type, with perhaps a tinge of blue, like the early
Korean wares' and that it 'was a beautiful ware of celadon type,
varying in tint from a very pale green to a bluish green'.[1] However, the final determination was left for Sir Percival David in his
valuable monograph on the subject.[2] This, in more ways than one,
constituted a landmark in Western studies of oriental ceramics,
not least for its critical appraisal of Chinese literary sources,
hitherto regarded as sacrosanct by Western scholars; however, on
the central point of identification, Sir Percival was doubtless
primarily indebted to the Chinese authorities at the Peking
Palace Museum, who had preserved the old tradition and had
access to a fine assortment of authenticated specimens in the
Collection. Several of these were displayed in the epoch-making
International Exhibition of Chinese Art held at Burlington House
in 1935-6 along with others which had been collected by Sir
Percival himself.

In his perspicacious remarks on Ju ware, Hobson had relied
heavily on the recorded passage by Hsü Ching, a Chinese scholar
who accompanied a Sung envoy to Korea in 1123, which stated
that the contemporary Korean celadons bore a general resemblance to the Five Dynasties Yüeh ware known as *pi-sê yao* and
'the new kiln wares of Ju-chou'. The similarity is indeed astonishing and indicates that there was a direct relationship between the
two wares. Since Korean celadon was developed under strong
Chinese influence, especially that of Yüeh ware, it is natural to
suppose that the Korean potters at a later date used Ju ware as their
model, finding its ease and grace specially to their taste. The result
has been that it is often easy to confuse the Korean porcelains with
the Chinese, and even after the most careful examination, there is
sometimes great difficulty in arriving at a correct attribution.
However, it has become evident in recent years that new styles
and techniques were freely copied by potters in both countries and

[1] R. L. Hobson, *Chinese Pottery and Porcelain*, (see Bibliography), Vol. I, pp.
42 & 54.
[2] Sir Percival David, *op. cit.*

were not an exclusively 'one way traffic', so that the possibility of reciprocal influence should by no means be dismissed. If the Korean potters were indebted to Ju ware for some of their finest productions, it is probable that the Ju craftsmen likewise seized on outstanding examples of Korean celadon to develop some of their own masterpieces of form and glaze. Indeed, the writer has seen a few specimens of Korean celadon, and has collected shards at Korean kiln sites, which are in no way inferior to the most highly praised examples of Ju ware, whether in their skilful potting, refined and elegant shapes or superb bluish glaze.

Returning to Northern Celadon, it is clear that this ware is a direct descendant of the Five Dynasties and early Sung Yüeh porcelain; and its relationship with another northern ware, Chün, also is close: indeed, one variety of the latter, known as 'Green Chün', is almost indistinguishable, though it is never enriched with the carved decoration which is such a feature of Northern Celadon. The depth and vigour of this lovely carved ornament, which usually takes the form of peony or other floral designs covering the entire surface of a vase or bowl, excel that of any other Chinese porcelain (Plate 11). The precision and sureness of touch are quite remarkable, and the whole achievement can only be described as masterly, the work of consummate craftsmen steeped in tradition. It is not surprising that Northern Celadon should have become so highly prized by collectors in recent years, and the prices obtained for top-ranking examples have mounted to very high figures. Thus, while Sir Percival David informed the writer that he purchased a large covered box with beautiful carved floral decoration for £10—probably in the early thirties—two small bowls freely carved with peonies from the Russell Collection were sold in 1960 for £1100 and £1200, and a slightly larger dish (7¼ in. in diameter) boldly carved with a peony spray from the Fuller Collection realized the astonishing figure of 7,800 guineas in June, 1965. These prices far exceed the sums given for the finest Lung-ch'üan bowls, the best price so far obtained for a typical conical lotus-petal bowl of high quality being £600 for one in the Norton Collection sold in 1963. It is evident that the feature which has so taken the fancy of collectors is the strongly carved decoration, which makes even a single piece of Northern Celadon stand out from a whole display of plainer celadon wares, the glazes of which may be of equal or even superior merit.

Whereas Northern Celadon has survived in abundance, examples of Tung ware are extremely rare—and indeed there is no assurance about the attributions which have been made (Plate 10). The description in the *Ko-ku Yao-lun* that 'it is of a pale green colour marked with fine crazing' holds good of most of the specimens proposed—and no doubt assisted in their provisional identification. Carved decoration of exceptional strength and depth seems to be characteristic, and the discovery of some examples of this type at the Yao-chou kiln sites (see page 50) lends colour to Sir Percival David's view that Tung ware may have been the precursor of Northern Celadon. There is in any case a close connection between the two wares, but the suggestion that Tung ware, or something like it, may have become the official ware after Ju (the *T'ao-shuo* also states that 'Tung porcelain resembles the imperial ware') perhaps indicates that the lines were always kept distinct. Frequent flooding of the K'ai-fêng area, where both Tung and Northern Kuan were made, has probably obliterated all trace of the kilns.

When the Chin Tartars invaded from the north and sacked the Sung capital early in 1127, they succeeded in capturing the reigning Emperor together with his illustrious father, Hui Tsung, who had abdicated in the previous year, and a large retinue. The remainder of the court were able to make good their flight southward, and a younger son of Hui Tsung was proclaimed Emperor, after escaping the general disaster and taking refuge in Nanking. However, this was not the end of their vicissitudes, for the Tartars later crossed the Yangtze and overran the provinces south of the river, capturing both Nanking and Hang-chou. The flight of the Sung court has, indeed, been called 'a complicated series of wanderings which filled eleven years'.[1] Eventually the Tartars were driven back by the reorganized Sung armies, and peace was concluded in 1141. After this, the Sung capital was established at Hang-chou, and it is recorded that official pottery kilns were set up at Hsiu-nei-ssǔ, the Imperial Household Department concerned with upkeep and repair of the palace buildings.

This demoralized flight to the south has been misconceived by some Western writers as a planned and orderly withdrawal, in which artists and potters followed in the wake of the court to its new seat at Hang-chou. Hobson referred to the official potters 'accompanying the court' and 'following in the imperial train',[2] while Palmgren stated that 'the emperor of the Sung dynasty, his family and court with all the higher dignitaries, with scribes and artists—even artisans—moved from Honan and the capital K'ai-fêng Fu across the Yang-tzu to Chêkiang and Hang-chou...(as) an emergency measure for the sake of security'.[3] This, as we have seen, is far removed from the facts, although we know that there was a general exodus to the south, and there can be little doubt that the principal northern pottery kilns fell on evil times or were discontinued and that many of the potters were sooner or later able to escape southwards.

Probably the new Kuan kilns were at first located within the palace precincts in the area reserved for workshops, accordingly coming under the same management, but it seems that they were in operation at this place for only a limited period, and it has so far proved impossible to locate the site or identify Hsiu-nei-ssǔ ware from shards and wasters. One reason for this is that the area was subsequently used for building, so that it is littered with fragments of the porcelains used by the palace staff or other residents of the villas erected there. The Chinese records state that it was a celadon (ch'ing) ware composed of fine material and exquisitely

[1] A. C. Moule, *The Rulers of China: 221 B.C.–A.D. 1949*, London, 1957, p. 88.
[2] R. L. Hobson, *op. cit.*, p. 89; also in *A Catalogue of Chinese Pottery in the Collection of Sir Percival David*, (See Bibliography), p. xix.
[3] Nils Palmgren, *Sung Sherds*, Stockholm, 1963, p. 1.

made; they praise its beautiful, translucent glaze and call it 'the delight of the age'.[1] There is a persistent theory in Japan that specimens of Sung celadon ware which have an almost white body and a pale bluish glaze devoid of crackle may be identified as Hsiu-nei-ssŭ ware; these are regarded in the West as superior Lung-ch'üan ware, and their claim to be Kuan rests mainly on their specially fine quality (Plate 14a & b). A somewhat similar conception to the Japanese seems to prevail among the custodians of the Palace Collection in Taiwan, except that most of the specimens so designated by them have irregular crazing and some of the best are held to be Northern Kuan. There are good grounds for believing that there was a close resemblance between Northern Kuan and at any rate the earlier, Hsiu-nei-ssŭ, Kuan wares made in the south, and Dr. Ch'ên of Peking strongly supports this view, which would make it extremely difficult to distinguish between the two. The position is further complicated by the fact that remarkably good copies of Kuan ware are known to have been made at other kilns, notably at Lung-ch'üan, where shards and wasters of this type were excavated in 1939 some twenty feet below the surface of the ground at Ao-ti (or Ao-li) near Ta-yao and at Tun-t'ou near Ta-ma (see sketch map on page 63).[2]

At some date which is not specified the Kuan kilns were transferred to a new site below Chiao-t'an, or the Altar of Heaven, and this has been reliably located at the foot of Tortoise Hill near Hang-chou. The official ware made at this place is said to have been inferior to the earlier product; nevertheless it is one of the finest of all Sung wares, characterized by a dark, often almost black, body very thinly potted and covered with numerous layers of glaze. Indeed, the glaze is often thicker than the body and shines with an opalescent lustre which is unmatched by any other type of Chinese porcelain; sometimes the effect is more subdued and resembles polished marble (Plate 12 & 13).

The Chiao-t'an site has been visited by many Chinese, Japanese

[1] *Fü-hsüan Tsa-lu, op. cit.*
[2] Wan-li Ch'ên, *op. cit.,* p. 28.

and Western scholars ever since its discovery in 1929-30 and a large number of fragments and wasters have been recovered, enabling the ware to be thoroughly examined and tested. The black colour of the body is attributable to a relatively large iron content fired in a reducing atmosphere, and the quality of the glaze to repeated applications—the layers are clearly discernible under magnification—and to the usual celadon features of unfused particles and minute bubbles held in suspension. But this by no means explains everything about the effect, which is another example of that remarkable flair possessed by the Chinese potter to produce empirically something which no amount of study or scientific knowledge can elucidate.

There is a good deal of variation in Chiao-t'an ware, and the modern potter can derive consolation from the fact that a number of pieces are light brown as a result of oxidation, which we must suppose to have been accidental, although this effect too is remarkably attractive. Crackle is almost invariable and of such beauty that it is generally assumed to have been contrived deliberately. Certainly it became a prized feature in many later Chinese wares, being produced and enhanced by a variety of artifices; but it is doubtful whether, at this early stage, the potters were able to exercise such precise control, although they certainly turned to advantage any chance effects which they observed—in the first place by using the materials and firing conditions found to give the desired results. Fragments and wasters do not show that the crackle has been accentuated, except where it has assumed a greyish tone due to impregnation with soil, while the darkening evident in perfect specimens could well have resulted from natural causes and was not necessarily produced artificially. Hetherington described this feature many years ago, which he noted was 'characteristic of some of the Sung wares especially of the Kuan type', in the following passage: 'They have brown craze lines running through them and are often so pronounced as to give the appearance of having been accentuated with pigment. The glaze in this case owes its slight colour to ferrous iron, and there is phosphate in the glaze. As a result, ferrous phosphate is present. Now if ferrous phosphate is exposed to the air, it will turn brown at atmospheric temperatures due to its oxidation, and the reaction is accelerated by sunlight. The craze lines furnish a means whereby the aerial oxygen can get contact with the ferrous phosphate and react with it. The aerial oxygen does not attack the phosphate in the uncrazed portions owing to the protection afforded by the surface skin of the glaze. The crackle, therefore, in this case is not due to the rubbing in brown colouring matter but by the natural oxidation of the ferrous phosphate at the glaze lines. This happening will occur in any phosphatic glaze, and some of the post Sung wares display the same effect, but it is particularly obvious in the Sung wares of this type.'[1]

One other feature of Southern Kuan ware should be mentioned, since it was much prized and talked about by Chinese connoisseurs and is repeatedly mentioned in literary sources, namely the 'brown mouth and iron foot' which were characteristic. The explanation given in the *Tsun-shêng Pa-chien* of 1591 is without doubt substantially correct and runs as follows:

'The reason for the brown mouth is that the liquid glaze being poured from above, ran down the side and covered the mouth less thickly than the rest of the vessel, so that at the mouth there were traces of the underlying brown. This is not in itself of any importance, but such points as "iron foot" are valued because no other place possesses such valuable clay.'[2]

The time has not yet come when it is possible to say anything useful about Ko ware, one of the five most famous Sung porcelains. The current view in the West is that this term, traditionally given to a special class of ware made by the elder of the two Chang brothers at Lung-ch'üan, later came to be associated with crackle

[1] A. L. Hetherington, 'The Why and Wherefore of Chinese Crackle', *Transactions of the Oriental Ceramic Society*, Vol. 15, 1937–8, pp. 19–20.
[2] See translation by Arthur Waley in *The Year Book of Oriental Art and Culture*, London, 1925, p. 84.

and to be applied generally to all types in which this was conspicuous, especially to a variety of Southern Kuan having a pale or slightly oxidized coloration. An article which appeared recently in a Chinese journal dealt mainly with the technical composition of the body and glaze material but without arriving at any final conclusions,[1] and the whole question needs further investigation. It may be premature to decide that there was no special class identifiable as Ko ware in the present state of our knowledge and in view of numerous references in the Chinese literature.

We now come to the most famous and best known of all celadon wares, the products of numerous kilns established in the vicinity of Lung-ch'üan in southern Chekiang. These were the celadons which arrived in Europe in the seventeenth century and excited so much admiration; but they had long been highly valued all over the Orient, being exported in vast quantities by sea to Egypt, Arabia, Persia, Africa, India and throughout the Far East. The Japanese, indeed, have made quite a cult of Lung-ch'üan ware, which they have divided into three groups according to the shade of colour, which in turn has depended largely on the period of make. Thus the finest bluish green celadons are known as *Kinuta*—signifying 'mallet' and derived from a celebrated mallet-shaped vase which possessed this admired tone[2]—and date from

[1] Jen Chou and Fu-k'ang Chang, 'Preliminary Studies of the Place of Manufacture of the Ko Yao Ware of the Sung Dynasty' (in Chinese), *Wen Wu*, No. 6, 1964, pp. 8–13, illus.

[2] The original *kinuta seiji* which gave rise to the generic name applied in Japan to all bluish green celadons is often said to be the mallet-shaped vase with phoenix handles or 'ears' originally owned by the Ashikaga Shōgun, Yoshimasa (1435–90), and now in the possession of the Bishamon-dō Temple in Kyōto. This celebrated vase was the finest of the three mallet-shaped vases displayed at the Japan Ceramic Society's Exhibition of Chinese Celadon Wares in October, 1950, and evoked much praise at the time for its 'heavenly blue-green glaze' (See the writer's article: 'Chinese Celadon in Tokyo', *Oriental Art*, Vol. III, No. 3, 1951, p. 124.). One is therefore at a loss to account for the comment attributed to Sir Percival David that it is 'a rough piece of Chinese celadon of poor quality' (Soame Jenyns, *Japanese Porcelain*, 1965, p. 11). In fact most of the *Kinuta* vases treasured in Japan are of much higher quality than any to be seen in Western collections, and this example is by general consent among the best. [*Continued on next page*].

the Sung period (Plates 15, 16, 17); the second variety is called *Tenryūji* and usually has a slightly olive tinge found mainly in celadons of the Yüan and early Ming period; while the third is known as *Shichikan*, made from about the middle of the Ming period and distinguished by a transparent, watery green glaze, often with a bluish tint. Lung-ch'üan wares were among the earliest used in Japan in connection with the Tea Ceremony in its original, aristocratic form at the end of the fifteenth century, and the majority of the finest mallet-shaped vases have been preserved from early times in Japanese collections.

It is not known when Lung-ch'üan ware was first produced, since there is a scarcity of references in early literary sources; however, thanks to the field-studies made in the past forty years and the survival of a few specimens of characteristic Northern Sung type, it is considered that the kilns were established during the Five Dynasties or early in the Sung period: the most valuable 'documents' found so far have been some pots of early type inscribed with the date 1080. Probably the Lung-ch'üan kilns developed and expanded after the close of the Five Dynasties period, when the Yüeh potteries were declining, and it is quite likely that some of the Yüeh potters moved southward and assisted in production of the new ware.

According to the legend repeated in Chinese literary sources, the rise of Lung-ch'üan ware is associated with the Chang family. The elder brother, Shêng-i, is said to have made Ko ware, characterized by a pronounced crackle, while the younger brother, Shêng-erh, produced the finer, smooth-glazed ware which rose to world-wide fame and set the standard for high-grade Lung-ch'üan

There is however, some doubt whether it is indeed the specimen to which the name *Kinuta* was first given. As in similar cases, there are different theories in Japan and several other claimants to the distinction, foremost among them being the mallet-shaped vase with fish handles formerly owned by the famous sixteenth-century Tea-master, Sen-no-Rikyū, which passed into the ownership of Baron Iwasaki. Another candidate is the vase of similar shape but without 'ears' and therefore more truly in the form of a mallet, which was long in the possession of the Daimyō of Kishu, a branch of the Tokugawa family. (See *Chinese Celadon Wares, op. cit.*, p. 51.)

celadon. Koyama recalls that exploitation of the Lung-ch'üan kiln sites greatly increased in 1916–17, when the area was thronged with antique dealers from Shanghai and Hang-chou and speculators acquired plots of land covered with kiln waste, which was carefully sifted and provided the large number of wasters, or imperfect pieces, sold on the market. Dr. Ch'ên visited the locality in 1928 and 1934 to carry out systematic investigations, collecting a great deal of material; and in recent years Chinese archaeologists have conducted a number of field-studies. The amount of work still to be carried out, however, must be enormous, since well over

LUNG-CH'ÜAN KILN SITES

two hundred kiln sites have been located and many others doubtless remain undiscovered.

The most famous kiln site is now known as Ta-yao, or the 'Great Kiln', and is popularly supposed to have been that of Chang the younger's pottery. The Swedish scholar, Dr. Palmgren, who visited the area in 1935–6, states that 'it is of gigantic dimensions and the sherd heaps cover large areas and form hills, nay even small mountains in the country. Most of the houses in the village are built upon the sherd heaps. Wherever you dig in the village, you will find sherds and still more sherds. Village traditions say that about 40 different kilns were active at the time in Liu-t'ien (the original name of Ta-yao). With the passage of time, the sherd heaps from the different factories have merged so that the inhabitants are no longer able to distinguish between different kiln sites. The deep ravine is the only dividing line.'[1]

Celadon wares which are heavily crackled or in other respects do not fit in with the current conception of Lung-ch'üan ware have often been called 'Hang-chou celadons', a practice which has been strongly criticized by the author of this book because there is at present no evidence of any local production in and around Hang-chou apart from that of the official, or imperial, Southern Kuan ware, and this is a different type altogether. Other kilns doubtless existed at Hang-chou, but we do not know their location or what types of wares they made. In this connection, it is of interest to note that some 30 per cent of the shards collected by Dr. Palmgren at the Ta-yao site had conspicuous crackle, while about 70 per cent were uncrackled: 'Crackled and uncrackled sherds occur in the same proportion on both sides of the Ta-yao canyon. The variations in the crackled Ta-yao ware are very prominent. The crackling effect depended partly on the thickness of the glaze (thick glaze cracks more easily), partly on the rapidity of the cooling. Uncrackled sherds generally have a glaze coating some 0·3–0·8 mm. thick, the crackled ones 1·0–1·5 mm. The crackling on the sherds in my material from Ta-yao is however to be considered a flaw in the manufacture.'[2] It should be added that the firing also determines the amount of crackle, as does the composition of the glaze; over-firing normally produces crackle, while under-firing results in a smooth surface without any cracks or crazing. Reducing the alkaline elements in the glaze and increasing the siliceous and boracic also tend to prevent crackle. It would, of course, be preferable always to employ the term 'crazing', as done in our discussion of Ju ware, since this feature should strictly be called 'crackle' only when it is produced deliberately.

[1] Nils Palmgren, *op. cit.*, p. 98.
[2] *Ibid.*, p. 104.

C. Wine bottle with pear-shaped body and tapering neck, flaring at the mouth. Korean. Koryŏ dynasty. Height 11.3 in. Mr. G. St. G. M. Gompertz.

Dr. Palmgren suggests as the most probable view that the ceramic requirements of the million or so inhabitants of Hang-chou were met by large-scale importation from the Lung-ch'üan potteries, especially from Ta-yao, rather than from local production. It seems in any case preferable to retain the designation 'Lung-ch'üan ware' in all cases except where another locality is named in an inscription or where the class of ware is so markedly different that it seems likely to have been made in another part of Chekiang Province. However, there is so much variation in the wares found at different sites near Lung-ch'üan that the description 'Lung-ch'üan ware' must be understood to cover a wider field than would normally be the case.

The younger Chang, we are told, 'improved upon the output of his elder brother in the fineness of his work and the excellence of his designs',[1] and this was doubtless due to the establishment of Hang-chou as the Southern Sung capital and the demand for fine ceramic wares by the court and well-to-do populace. Both Dr. Ch'ên and Dr. Palmgren have confirmed that the shards from Ta-yao were of the highest quality and included many of the *Kinuta* class as well as others bearing inscriptions, mostly in seal-characters. As production increased, it seems that the kilns spread northwards and eastwards along both sides of the River Ta-ch'i, and much of the transportation was doubtless effected by means of boats and rafts floated down the river. Ta-yao is situated about thirty miles south of Lung-ch'üan, and there are numerous kiln sites further to the west as well as down-river to a distance of some thirty miles east of Lung-ch'üan. Many of these continued active in the Ming period, and in a brief summary Dr. Ch'ên shows 6 kilns active in the Sung period, 6 from Sung to Ming, 3 from Yüan to Ming, 10 Ming, 1 Ming to Ch'ing and 6 Ch'ing. This effectively disposes of the report that the kilns were all moved from Lung-ch'üan to the Ch'u-chou district, near Li-shui, at the beginning of the Ming period (Dr.

---

[1] According to the *Ch'ing-pi Tsang*, published in 1595, quoted in the *T'ao shuo*.

Ch'ên roundly states that 'this is not a historical fact'), which has sometimes caused later celadons of Lung-ch'üan type to be described as 'Ch'u-chou ware'. However, there is a tradition that 36 kilns near Ta-yao were destroyed by flood, and Dr. Ch'ên found evidence that the kilns in this area had been abandoned simultaneously as a result of some misfortune.[1]

Although Lung-ch'üan ware was mass-produced, the scale of production was so large that there was a profusion of different shapes. At first, however, decoration was severely restrained, if not altogether absent; the potters relied on the simplicity and purity of the forms and the beauty of the blue-green glaze, just as did the makers of Imperial Ju. In the late Sung and still more in the Yüan and Ming periods a great deal of incised, carved or impressed ornament was introduced; but at the same time there was a loss of quality in the glaze, which began to develop a faint olive tone or else was a fine but unvaried 'sea-green' colour. The decoration was often stressed by leaving moulded fish or dragons inside dishes bare of glaze, with the result that they turned a rich brown in the firing (Plate 19), or by applying spots and splashes of iron oxide, producing the variety of ware known in Japan as *tobi seiji*, or 'spotted celadon', with brown markings (Plate 18). Koyama attributes the decline in quality to increased popularity and consequent over-production; and there can be little question that the ever-increasing scale of production and export made it difficult to maintain the original high standards. Some 1300 celadons, mainly dating from the Ming period, are still preserved in the Topkapu Palace at Istanbul, while there are 58 late Yüan and early Ming celadons from the Ardebil Shrine (near the Caspian Sea) at Teheran. A characteristic of these post-Sung celadons from Lung-ch'üan is that most of them are relatively large and heavy dishes, bowls and vases of a type that must have been well suited to transportation over long distances: their robust construction minimized damage, and their size no doubt excited admiration and facili-

[1]Wan-li Ch'ên, 'Second Report on the Examination of the Lung-ch'üan Kiln Sites' (in Japanese), *Tōji*, Vol. VII, No. 5, October 1935, p. 2.

tated display in the large rooms and palaces common to the Middle East (Plate 20).

The body of Lung-ch'üan ware is greyish white in colour but, like other celadons, has turned reddish brown wherever it was exposed to the firing without a protective covering of glaze. This is particularly evident at the foot, where the glaze has ended, so that it is said to have 'turned red in the firing'. Accentuation of this feature at the termination of the glaze is probably due to the increased amount of iron present, derived from both glaze and body, and is thought to have taken place at the end of the firing, when air is admitted into the kiln. Thus, as Hetherington remarked, it would be more correct to say that the body 'has turned red in the cooling', since the strongly reducing atmosphere present during the actual firing has eliminated the oxygen and resulted in the iron present being converted into ferrous oxide, producing the cool bluish green tones which are characteristic of the ware.

During the Ch'ing period celadons were made at Ching-tê-chên with a pure white body and, beautiful as are many of the shapes, the whole character of the ware seems to have changed, bringing about a loss of individuality. The glazes are often attractive and possess a refinement which imparts distinction, but technical proficiency is not a compensation for loss of vitality and inspiration. At the great potteries of Ching-tê-chên finesse and sophistication reached a high point, together with an archaistic tendency which resulted in some incredibly fine imitations of the early Ju and Kuan porcelains. It is sometimes very difficult to distinguish the best of these from their original models, but like all copies they seem to have gained in superficial attractiveness at the expense of creative power. Some weakness must always be inherent in an imitation, however accomplished its execution. However, the most experienced collectors have often been deceived, not least the Emperor Ch'ien-lung, who often mistook copies for originals and caused misleading inscriptions to be incised on their bases with a high degree of calligraphic skill but to the utter confusion of modern students. The practice of such ingenious artifices as burying specimens in 'filthy sewers' or immersing them in strong solutions and 'stewing' them in the concoction, as mentioned by Père d'Entrecolles, who visited the Ching-tê-chên potteries early in the eighteenth century, imparted a deceptive appearance of age and wear, so that a great deal of experience and discernment is necessary to enable a correct attribution to be made. Such specimens call for a supreme test of instinct and judgment on the part of the connoisseur, for there is no ready guide to a correct appraisal, although scientific examination of the sub-surface structure of glazes may offer a means to check doubtful cases.[1]

Dr. Ch'ên mentions a jar believed to have been made in the Yung-lo era (1403–24), the colour of which 'rivals that of the finest Lung-ch'üan products', while the copies of Ju, Kuan, Ko and Lung-ch'üan made in the Yung-chêng (1723–35) and Ch'ien-lung (1736–95) eras 'are so fine that they are almost indistinguishable from the Sung wares'.[2] In modern times Japanese potters have excelled in their imitations of Lung-ch'üan ware, and at least one Kyōto potter is said to have made copies of wasters which show as much character and individuality as authentic specimens, something which may well baffle the most expert, for it would seem almost inconceivable that anyone would deliberately imitate wares which have been spoiled by some mishap.

Another use of celadon glaze in some Ch'ing polychrome porcelains had nothing to do with the copying of early wares. It was employed either as the ground colour or for part of a design. Such motifs as birds among prunus branches might be in underglaze blue, while the background was celadon—pale olive or sea-green in hue and transparent enough to reveal the white body underneath. In the eighteenth century French goldsmiths often provided wares of this kind, also the celadon monochromes, with elaborate ormolu mounts—a tribute to the high esteem in which they were held.

[1] See Robert T. Paine, Jr. and William J. Young, 'A Preliminary Report on the Sub-surface Structure of Glazes of Kuan and Kuan-type Wares', *Far Eastern Ceramic Bulletin*, Vol. 5, No. 3, September 1953.

[2] Wan-li Ch'ên, *op. cit.* (*Chung-kuo Ch'ing-tz'u Shih-lüeh*), p. 45.

CHAPTER FOUR

Korean Celadon

The potter's art has a long history in Korea, for there is something in the make-up of the Korean which leads to his becom-
ing a natural craftsman, perhaps not as technically proficient as the
Chinese, but even more spontaneous and artistic. No-one has
written with greater understanding of this Korean trait than
Sōetsu Yanagi, who was a pioneer in the appreciation of Korean
art-works, and it is worth quoting the following passage on the
'nonchalance' of the Korean artisan and his consequent freedom
from artificial restrictions, whether in turning wood or throwing
pots:

'Their way of making things is so natural that any man-made
rule becomes meaningless. They have neither attachment to
the perfect piece nor to the imperfect . . . Their state of mind is
free from dualistic, man-made rules . . . Here, deeply buried, is
the mystery of the endless beauty of Korean wares. They just
make what they make without any pretension . . . One who has
the chance of visiting a Korean pottery [Dr. Yanagi is referring
to a contemporary pottery, but his remarks have a general
application—Ed.] may notice that the wheel used for throwing
pots is never exactly true. Sometimes it is so crudely set up that
it is not even horizontal. The asymmetrical nature of Korean
pots comes partially from the uneven movement of the surface
of the wheels. But we must understand that Koreans do not make
such wheels because they like unevenness, but they just make
their wheels in that happy-go-lucky way . . . Of course, if the
wheel slopes too much, they may correct it to some extent, but

that will not mean precision even then. They hardly trouble about accuracy or inaccuracy. They live in a world where accuracy and inaccuracy are not yet differentiated. This state of mind is the very foundation from which the beauty of the Korean pots flows out . . . '[1]

The Koreans had for several centuries made unglazed earthenware and stoneware, but it was not until the end of the tenth century that they began experimenting with high-fired stoneware covered with feldspathic glazes, and there is abundant evidence, both from the shapes of the vessels and the technical details as well as the finely incised decoration which was later introduced, that they were indebted to the makers of Yüeh ware for the development of their craft. We have seen that the state of Wu-Yüeh was swallowed up by the Sung empire in the year 978, after which the potteries lost their princely patronage and entered on a slow decline. While there is no direct evidence, the indications are quite strong that some of the Chinese potters may have emigrated and settled in southern Korea, and it is noteworthy that the first large pottery centre in that country was established about ten miles south of the town of Kangjin, at the southwestern extremity of the peninsula, or the nearest point to the Yüeh potteries in Chekiang. The normal trade-route between China and Korea at that time was by the short sea passage across the Yellow Sea after an easy coasting trip from such ports as Ning-po to a point opposite the southern tip of Korea; the journey to the Korean capital of Songdo, the modern Kaesŏng, could then be continued, if necessary, along the Korean coastline, sheltered by innumerable bays and islands. Both Wu-Yüeh and Korea were strongly Buddhist states, and there must have been continuous traffic of monks and sacred articles as well as merchandise by this means.

The earliest kilns were established at the head of a steep valley leading from a small Buddhist temple down to a small bay, which

---

[1] Sōetsu Yanagi, *A Harvest of Folk-crafts*, Tōkyō, 1963, pp. 14–15. (Vol. III of *Illustrated Cyclopedia of Folk-crafts*).

was situated on the eastern shore of a wide inlet of the sea, pro-
tected by mountains and islands. As time went on, fresh kilns were
constructed lower down, and this process continued until, by the
first half of the twelfth century, the kilns had spread downwards
almost to the sea-shore. It is a place of great natural beauty and
must have presented a busy scene eight hundred years ago, with
porters and ox-carts carrying their loads of finished wares down
to the water's edge to be shipped by sampans and small junks up
the island-studded west coast to harbours near the capital.

On the north bank of a rocky stream running down the valley
there are some cultivated fields on small levelled areas, where such
crops as millet and sweet-potatoes are now grown, and among the
rows of produce are numerous celadon shards, where the soil has
been turned over in the course of farming. On the south bank the
ground rises more steeply; there are outcrops of rock, and the
terrain is so rough that it has never come under cultivation. Here,
among the jagged rocks, there are great masses of kiln waste,
lumps of fused clay and saggars, often with broken segments of
celadon bowls still adhering to them. All the fragments found in
this area come from the earliest wares, roughly potted and covered
with a dull, grey-green glaze or light brown as a result of oxidation.
All the characteristic Yüeh features are in evidence, especially
with regard to the bases and spur-marks, for the kilns were
active in the period when celadon ware was still being perfected.

Lower down, however, only a few hundred yards from the
sea-shore, are the sites of the later, twelfth-century kilns, especially
the ones at Sadang-ni, where the finest wares were produced; and
here, only a few weeks before the writer's visit in October 1964,
one of the greatest pottery finds of all time had been made. As long
ago as 1914, when the Kangjin kiln centre was first discovered,
part of a celadon roof-tile had been discovered in this area, and in
1928 another very small fragment bearing relief decoration was
picked up, while several other pieces were collected at the site of
the royal palace in Songdo. The importance of these discoveries
derives from the fact that the History of Koryŏ—the name of the

Korean state from A.D. 918–1392—records a special commission by the King for celadon tiles in the year 1157, which were required for roofing an ornamental pavilion at the site of the palace. Since there is no other record of celadon ware being used for such a purpose, the indications are quite strong that these fragments came from tiles manufactured for this royal pavilion, and the date of their make can accordingly be precisely fixed at the middle of the twelfth century, when the techniques of glazing and producing superior designs in relief are considered to have reached their zenith.[1] No other finds of this kind were made until the National Museum of Korea carried out a survey just before the writer's visit. From various indications it was decided that excavations within the mud-wall bounding a small cottage at Sadang-ni might prove fruitful, and the results were spectacular, for this proved to be the site of a hitherto undiscovered kiln which had evidently made the roof-tiles. A great quantity of fragments and wasters were recovered, including some half-dozen almost complete roof-tiles, with decorated faces some ten inches in length and three inches wide, together with the circular face of a 'round' roof-tile. The quality of the celadon was of the best, a beautiful bluish green in colour, and the relief decoration of floral arabesques or medallions was the finest ever seen. Here at last was indisputable proof that the fragments recovered earlier were in fact pieces of roof-tile and that the celadon tiles for the royal pavilion had been manufactured at the Kangjin kiln centre.[2]

The discovery of a second large pottery centre in 1929 at Puan, on the west coast of Korea some eighty miles north of Kangjin, provided further stimulus for the investigation and study of Koryŏ celadon. The writer visited this site also in October 1964, paying chief attention to the Yuch'ŏn-ni section, which comprises

[1] For a fuller account of these discoveries see the author's book: *Korean Celadon and Other Wares of the Koryŏ Period*, (see Bibliography), Chapter 5: 'Celadon Roof-tiles for a Royal Pavilion'.

[2] Since writing this, word has been received that further excavations have taken place and resulted in the recovery of enough tile wasters to enable part of a roof to be reconstructed.

a wide area of undulating land now under cultivation, except where it is broken up by hillocks, rough ground or scrub, beyond which a great arm of the sea extends inland. Here, in the twelfth century, were established numerous kilns which made celadon wares of high quality, in some cases perhaps even finer than those produced at Kangjin; and to this day the ground is studded with fragments of celadon, shining in the sunlight like blue and green gems. It is a thrilling experience to have this vision of a great artistic achievement and, indeed, of a considerable industry, whose relics, however fragmentary, have persisted through so many centuries with such eye-catching beauty.

A few general observations on the shards collected will not be out of place at this point. The first and most striking point is how *blue* many of the fragments appear—but this should be understood to mean a celadon rather than a Chün blue. As already mentioned, Orientals do not distinguish clearly between blue and green, and perhaps this is the wisest course, though it has led the Chinese to become involved in such equivocal descriptions as 'sky-clearing blue' and 'cucumber-green'. But there is no gainsaying that many of the fragments, as seen in bright sunshine, are of a truly ethereal blue, like bits of the azure sky fallen to the ground, and this impression is not dispelled by examination under less favourable conditions of light, after the shards have been washed and cleaned. Secondly, it is evident from the very first that there are a great number of different shades and tones of colour, so that on later examination one is in some doubt whether to arrange the shards in accordance with their type and decoration—for they include every kind of design, incised, carved, impressed and inlaid as well as plain, undecorated pieces—or simply by their colour. Finally, it is surprising how many unglazed fragments are picked up, but whether this is an indication that there had been a preliminary low-temperature firing—a point not yet fully determined with regard to celadon ware—or whether it is merely the result of kiln mishaps cannot readily be decided. Besides the fragments of celadon, there are numbers of the round pads of earth or clay used as supports during the firing, together with pieces of raw quartz or silica, the abundance of which must have prompted its use as spurs or stilts on which to stand the celadons during the same operation. No trace of the actual kilns could be detected, nor were any complete saggars to be seen, though some of the bases of bowls had portions of saggar adhering to them, but surface remains of the kilns together with much waste material has doubtless been obliterated by centuries of cultivation. However, the profusion of shards in certain places was very marked and must have indicated the proximity of particular kilns.

A few other isolated groups of Koryŏ period pottery kilns have been located in other parts of southern and western Korea, one as far north as P'yŏngyang, and there is a tradition that there were kilns at Songdo itself; but none of these approaches in size and importance the two great centres we have described. Late in the period, that is to say in the thirteenth and fourteenth centuries, it seems that many additional kilns were constructed: several have been discovered in recent years, for example, on the foothills of Mudung-san, a mountain just east of Kwangju, but by this time there had been a decline in the quality of the ware, and production was probably on a larger scale and no longer limited to providing the needs of the court and aristocracy. The same may well be true of the other groups mentioned, but investigations have not yet been carried to the point where their products can be accurately dated.

The record of Hsü Ching, who it will be recalled was the Chinese scholar accompanying the Sung envoy to Korea in 1123, has been carefully analyzed and discussed so far as it refers to ceramic wares.[1] Hsü Ching has proved to be an acute observer, and his remarks, though brief, provide an invaluable source of information on the celadons of the time and their relation to contemporary Chinese products. The Chinese, of course, had a

[1] See the author's study: 'Hsü Ching's Visit to Korea in 1123' in the *Transactions of the Oriental Ceramic Society, 1960–2,* Vol. 33, reprinted in slightly abbreviated form as Chapter 3 of *Korean Celadon and Other Wares of the Koryŏ Period* (see Bibliography).

natural arrogance and were not disposed to accord extravagant praise to the rival products of outlying kingdoms, which they regarded as semi-barbarous, and Hsü Ching implies that most of the Korean wares were no more than copies of Chinese porcelains. It is all the more remarkable that he was so obviously struck by the beauty of the Korean celadon glaze, which he notes had shown great improvement in recent years, and by the novelty and in-genuity of some of their ceramic forms (Plate 22). When these reactions are combined with another record, also dating from the Sung period, which lists Korean celadons along with the porce-lains of Ting-chou as 'first under Heaven' and furthermore makes use of the hallowed term *pi-sê* ('secret' or 'reserved colour') nor-mally applied to the highest class of Yüeh ware to describe them,[1] we can see that the Chinese were very much impressed by Korean celadon ware, which they clearly held in high esteem. As sug-gested in our comments on Imperial Ju ware (see pages 54–5), it is not improbable that the Korean celadons exercised an influence on the Chinese ware hardly less than that which they had themselves gained from Yüeh, Ting, Ju and other Chinese products; but this is a question which requires much more investigation and can only be resolved after further discoveries by archaeologists and literary scholars.

The survival of many hundreds, perhaps thousands, of Koryŏ wares, nearly all of which have been recovered from old tombs, has proved beyond all question that Korean celadon represents one of the summits of ceramic achievement. The glaze was in-finitely varied—far more so even than that of Lung-ch'üan ware—and at times reached a standard which, as Hobson remarked, 'even the best Chinese potters might have envied'.[2] The late W. B. Honey, former Keeper of the Department of Ceramics at the Victoria and Albert Museum, described the Koryŏ wares as 'some of the most beautiful pottery ever made' and referred especially

[1] The passage occurs in the *Hsiu-chung-chin* by the Sung writer T'ai-p'ing Lao-jên.
[2] R. L. Hobson, 'Corean Pottery—II The Koryu Period', *The Burlington Magazine*, Vol. LVI, April 1930, p. 193.

to their simplicity and beauty of form, their unaffected gracious-
ness and dignity. They seem, he said, 'to speak of a serenely happy
people' and are inferior to Chinese wares only in the range of
technical resources employed.[1] Among the Japanese, Koyama has
gone on record that 'there are no wares so beautiful as the
Korean—this is a concept quite widely held among us' and has
referred to their 'quietness' and subtlety.[2] In the words of Sōetsu
Yanagi, they 'wait for you' and are the most satisfying of all
pottery to live with, never making any assault on the senses or
'disturbing one's feelings'. Such praise is high indeed, coming
from acknowledged experts and connoisseurs of ceramic art.

It is time to give some details of the practical achievements of
the Koryŏ potters, after which we shall describe the main features
of their celadon glaze. The forms of their wares were perhaps
somewhat limited, comprising mainly dishes and bowls of various
sizes, wine-cups and stands (Plate 29), covered boxes—often with
sets of smaller boxes nested inside—wine-pots (Plate 27), ewers
(Plate 28), bottles (Plate 26) and prunus vases; but they also made
incense burners with covers surmounted by animals and birds
modelled in the round (Plate 22) as well as water-droppers likewise in the form of animals (Plate 24a), birds and human figures,
especially serving-boys. These ceramic sculptures are quite extra-
ordinary in their virtuosity and a certain baroque elegance which
contrasts strangely with the simplicity and restraint shown in most
of their pottery forms. Their greatest contribution to ceramic art,
however, was the technical device of inlaying with different
coloured clays to produce strikingly contrasted decoration. This
method was employed very rarely in China and never exploited
or carried to the same perfection, so that we are justified in calling
it a Korean invention. The decorative patterns were first incised in
the soft body of the vessel, after which reddish brown and white

[1] W. B. Honey, 'Corean Wares of the Koryu Period', *Transactions of the
Oriental Ceramic Society*, 1946–7, Vol. 22, p. 9; see also the same writer's book,
Corean Pottery (see Bibliography).

[2] Fujio Koyama, *Tōki Zuroku* (Illustrated Catalogue of Ceramics), Vol. 9,
p. 1.

slips were carefully brushed or pressed into the incisions and then, after the surface had been cleaned or shaved to remove any surplus slip, the glaze was applied and the vessel fired in the usual way. It is uncertain whether any preliminary low-temperature firing was carried out immediately after the inlaying process and prior to the application of glaze. After completion, the designs stood out in black and white, making a pleasing contrast with the green background (Plates 29 & 30). At first the quality of the inlaid wares was equal in every respect to those on which designs had been incised or impressed, and the glaze was no less fine in colour and lustre; but later it seems the new style became so popular that it caused over-much attention to be devoted to the decoration, with the result that less care was given to production of a superior glaze, and the colour often became greyish green or uneven. Finally, with the decline in prosperity consequent upon the Mongol invasion, which began in 1231 and subjected the whole country to ruin and devastation, the quality of the wares suffered further impairment and the potting became coarse and heavy, the decoration commonplace and the colour dull or brownish. The period of peak activity and superior quality started about the middle of the twelfth century but had already begun to wane by the time of the first Mongol incursions and, after a period of renewed but fading splendour during the latter part of the thirteenth century, the wares showed a further sharp deterioration, so that in the fourteenth century the forms had become extravagant, the decoration pretentious and the colour debased.

Inlaid decoration at its best is uniquely suited to the graceful elegance and good taste which inform Korean ceramic art and enables the native sensibility to be expressed with great freedom and strength. The floral patterns are often as delicate as fine lacework, but the finest examples are seen in the freely-drawn designs of cranes flying among clouds and ducks swimming among reeds and lotuses, which were portrayed with the flowing lines and subtle curves typical of the best Korean art.

Another technical invention which there is good reason to

credit to the Korean potters is the use of copper oxide, painted under the glaze and producing bright red decoration when fired in a reducing atmosphere. This seems to have been a very tricky process, for both the Korean and Chinese potters encountered many problems in bringing it to a successful outcome. In fact, it was not until the seventeenth century that effective control was fully assured. The earliest Korean celadons decorated with copper-red are stylistically datable to the latter half of the twelfth century, and this view has been supported by the discovery of shards at the kiln sites mixed with other fragments made at about this time. The invasion and occupation of Korea by the Mongols resulted in close ties being formed with China, which came under Mongol rule in 1280, and it is likely that the use of underglaze copper was introduced from Korea about the same time, since the earliest porcelains with this type of ornament are believed to date from the late thirteenth or fourteenth century.

We have seen that Hsü Ching was impressed with the beauty of the Korean celadon glaze, and some remarks on this outstanding feature will be desirable by way of conclusion. Hsü Ching re-marked that the Koreans called their wares 'kingfisher coloured', and this seems to have been either a play on words or a misinter-pretation of the Chinese name *pi-sê* ('secret' or 'reserved colour') applied to the highest class of Yüeh celadon, since the word for kingfisher has the same pronunciation in Korean, although the character is of course different. This term, incidentally, came into use to describe the dark green jade of Sung times, so that the precise interpretation is uncertain, and it could equally well have stood for 'jade-coloured'.

The chief feature of the Korean celadon glaze is that it was very thin and watery, being almost transparent where present only as a light coating but a milky green where it has run thick, especially when slightly under-fired. This resulted in the glaze running off the raised parts of a relief design and pooling in the depressions round its edges, which has imparted a striking effect of light and shade to impressed decoration. It should once more be stressed that the grey body of the ware and the conditions under which the firing has taken place are vital factors in determining the colour and lustre of the glaze, so that there is a remarkable amount of variation from bluish green and emerald to grey-green and dove-grey. This has given great individuality and character to Korean celadon, more so perhaps than to any one of the Chinese celadon wares; and it is very evident among shards collected at the kiln sites, which can be grouped into series of different shades. The Korean celadon glaze has with good reason been compared with sea-water, which displays so many varied hues according to the conditions of the light, the passage of clouds across the sky and the depth and consistency of the water. In contrast to the glaze, the body is composed of an almost uniform greyish coloured clay, but this, too, is ferruginous and sometimes changes colour in the firing, affecting the tone of the overlying glaze.

Enough has been said to show that Korean celadon is one of the major achievements in oriental ceramics. At their best, the Korean celadons are unsurpassed, whether in form, style or quality of glaze; but they are less well known than the Chinese celadon wares, and most of the finest pieces can only be seen in Korea and Japan. Collectors are apt to rate ceramic wares in direct proportion to their ability to gain possession of them, and good quality Korean celadons have always been scarce in the West. Again, the most distinctive of Korean celadon wares are those which have inlaid decoration, and the earlier inlaid wares with glazes comparable with the best of the other varieties are extremely rare, while Western connoisseurs are still relatively unfamiliar with this style of decoration and therefore disinclined to pay the high prices necessary for their acquisition. Indeed, it is only quite recently that Western scholars have come to realize the high quality of Korean celadon, for hitherto Chinese celadon has always been the acknowledged leader in this field. The limited number of second-rate Korean wares which have found their way into the hands of dealers only deepened a conviction that there was nothing here which need distract anyone from appreciation of the

Chinese wares. But first, somewhat diffidently, Hobson, and then, with more force and discernment, Honey, put into writing what they had discovered about Korean celadon, with the result that Western students no longer have any excuse for illusions on the matter. Unfortunately, it cannot be anticipated that any increasing number of Korean wares will become available for purchase. Korean graves continue to be despoiled—a disaster for archaeological research, for not more than half-a-dozen controlled excavations have ever taken place and nearly all the articles recovered from tombs of the Koryŏ period have come onto the market without any record concerning their place or manner of discovery. However, there can be little doubt that the royal tombs near Kaesŏng have all been ransacked, and only a few further finds can be expected from time to time as other tombs come to be located in different parts of the country. In fact, there are probably more Korean celadons secreted in private Japanese collections and remaining unpublished than we are likely to see as a result of the pillaging of tombs in Korea itself.

The earliest recorded tomb-robbery took place in the winter of 1884–5 in the vicinity of Kaesŏng, and some of the pieces found were acquired by one of the first British consular representatives in Korea.[1] After this, however, no further depredations seem to have occurred until after the Russo-Japanese War, when an era of road and railway construction opened and many celadon wares were accidentally dug up. The demand which then arose from collectors resulted in steadily mounting prices being paid for the wares, and thousands of Korean peasants are said to have made a living by surreptitious excavations, using long iron probes to locate underground tomb-chambers. This resulted in many of the pieces recovered being fractured or otherwise damaged, and no records are available concerning their place of discovery or what other articles were found with them. Most of the specimens acquired by Western collectors and museums, of which the best are in America, were excavated in the 'twenties and 'thirties, but

[1] See W. R. Carles, *Life in Corea*, London, 1888, pp. 139–41.

the great majority found their way into Japanese collections, since there were some half-million Japanese living in Korea during the period of Japanese rule and the Japanese have a strong bent for antiquarian studies as well as being avid collectors of antiques and *objets d'art*, especially ceramics. Unfortunately the government authorities and museum experts were preoccupied with the preservation of ancient monuments and with archaeological researches at P'yŏngyang, the site of a Han Chinese colony from 108 B.C. until the fourth century, and in the Kyŏngju region, where the Korean capital was situated until early in the tenth century. Little opportunity could accordingly be found for investigating Koryŏ period tombs and kiln sites, with the result that the materials available for studying the development of celadon ware are scanty in the extreme. It has, indeed, been possible to work out a reasoned chronology only on the basis of scattered clues and serial studies, which have enabled the general trend to be established, but with very little in the way of precise dating to prove beyond serious question just when the various stages of development took place.

CHAPTER FIVE

Japanese Celadon

We saw earlier (page 43) that the first great Chinese export ware was Yüeh celadon and that specimens or fragments dating from the T'ang period have been recovered from many parts of the Middle East as well as the Far East. Among these were some important examples discovered in southern Japan, the earliest being an incense-jar preserved from the eighth century among the treasures of Hōryū-ji Temple, near Nara. There are also references in early Japanese literature, notably in the 'Tales of Genji', to 'celadon' (*aoji* and *seiji*) and 'secret colour ware' (*hishoku*), showing that Yüeh ware was known and highly prized under the Fujiwaras, or in the tenth and eleventh centuries. By the twelfth century Korean celadons also had risen to fame, and a few specimens of Koryŏ inlaid celadon have been preserved in Japan from the time of their arrival, probably in the thirteenth century. It is not surprising therefore that the earliest glazed stoneware made in Japan attempted to reproduce these famous continental wares and was in fact a type of celadon.

According to the Japanese records, the introduction of a celadon glaze took place in the Heian or early Fujiwara period, perhaps as early as the ninth century. At this stage in the development of Japanese ceramics the glazes employed were derived from natural wood-ash, and we have no means of knowing how successful were the copies of Chinese and Korean celadons, since no examples of the 'Owari *seiji*' mentioned in contemporary records are known to have survived. Owari Province (now Aichi Prefecture) was a leading centre for the production of Sue ware, which had been derived from Korea about the fifth century and involved the

82

relatively advanced technique of throwing on a wheel and firing in a kiln; so it was natural that further steps in the introduction of continental methods should take place in that region.

By the twelfth century the centre for ceramic production of this type had been established in the vicinity of Seto, and the most important Japanese ware made in the Kamakura period (1185–1333) has come to be known as Ko-Seto, or 'Old Seto'. Although a true celadon ware, the early Seto pieces which have survived show the effect of firing in an oxidizing atmosphere: the glaze colour is usually amber or yellowish, often called 'dead-leaf' or 'caramel' colour by the Japanese. The fact is that the Seto potters had set themselves a task of great difficulty: 'the interesting thing is not how far short they fell from their mark, but rather that in the process they were able to utilize their own resources to produce wares of great appeal in their own right, and in many ways fully as creditable as the great Sung celadons upon which they were originally modeled'.[1] Their technique was inexpert and the materials available in Japan could not match those used in making Chinese and Korean celadons. Although the glaze contained feldspar and the firing was effected at a high temperature, Seto ware is much inferior to the celadon then being produced on the continent. The glaze is usually uneven and streaky, rarely attaining the lustrous green which was no doubt the potters' aim. To some extent these shortcomings were overcome from the early Muromachi period, or after 1333, when the glaze contained more feldspar and produced a more stable, smooth and even surface.

Some two hundred kiln sites dating from the Kamakura and early Muromachi (or Ashikaga) period have been found among the wooded hills surrounding Seto town, all strewn with shards covered with the characteristic Old Seto glazes, which included a *temmoku* black as well as the amber or olive-green. These were mostly from vessels made for Buddhist or Shinto ritual use, comprising jars with loop-handles, vessels for rice-wine in *mei-p'ing*

[1] Roy Andrew Miller, *Japanese Ceramics*, Tōkyō, 1960, pp. 33–4.

shape, ewers, flower-vases, incense burners and guardian lion-dogs. The influence of the continental wares is clearly evident from the decoration incised, carved, stamped, impressed or applied in relief. An interesting study of this was made some years ago by Sensaku Nakagawa and showed the use of typical Lung-ch'üan, *ch'ing-pai* and Koryŏ motifs, the pendent willows and chrysanthemum arabesques drawn from the Korean celadons being specially noteworthy.[1]

Unfortunately the study of the Old Seto wares has been impeded and bedevilled by the scandal which took place in recent years when some of the best known examples, including at least one with a dated inscription which had been accepted as genuine by the highest authorities, were shown to have been made as a prank by a contemporary Seto potter, who obligingly produced some more identical specimens to order. There can be little question that the number of spurious Seto wares represents only a small proportion of those in existence; however, until further investigation has resulted in some reliable touchstone being discovered, no-one can be one hundred per cent sure that any but a few excavated or otherwise authenticated pieces are 'right'. The example here illustrated is one of these unimpeachable specimens, and furthermore it is a rarity in having a fine transparent green glaze (Plate 31). The writer is indebted to Colonel J. G. Figgess of the British Embassy, Tōkyō, for this photograph together with the interesting data which follow: the wide-mouthed jar contained the cremated ashes of a Zen priest named Shin'e Chikai Risshi and was excavated from his tomb in the precincts of Kakuon-ji Temple at Kamakura, the tomb being marked by a stone pagoda. The excavation was undertaken in January, 1965, because it became necessary to dismantle and repair the pagoda. It is known from documentary records that Shin'e died in 1306 and that his ashes were placed in a jar. The stone pagoda was erected in 1332, and the jar containing the ashes was buried under

[1] Sensaku Nakagawa, 'Designs on "Old Seto" Porcelain', *Bijutsu Kenkyū*, No. 201, November, 1958, pp. 125–36, illus.

the foundations. When this jar was recovered during the excavation, the ashes were accordingly transferred to a new vessel and the original jar placed on display as an authentic specimen of Kamakura Seto ware.

It is surprising that the Japanese, having thus initiated the production of high-fired porcellanous ware with a type closely modelled on the continental celadons, thereafter proceeded to make a variety of wares with white, black or other glazes and never reverted to celadon except on a relatively small scale and usually in imitation of Chinese wares. Individual potters made celadons from time to time with great success, generally in the style of the Ch'ing period wares, that is by applying a celadon glaze to a white-bodied porcelain, but none of their wares became an accepted Japanese type or continued in vogue for any length of time. Yet Japanese technical skill by now was fully equal to the task: there are Kakiemon and Nabeshima celadons of the highest quality and distinction, while some of the most deceptive copies of Lung-ch'üan ware have been made by Japanese potters. The traditional wares, however, all developed along different lines.

About the year 1615 the potter Takahara Goroshichi, who is said to have been of Korean or Chinese extraction like many others at this period, made celadon ware in the vicinity of Arita and imparted the technique to the first Kakiemon and to Soyada Kuzaye-mon, who became chief potter at the Iwayagawachi kiln, established by the Nabeshima family in 1628. Fragments of celadon have been reported from this site, which is close to Arita. Later in the century the Nabeshima kilns were moved a few miles north to the famous Okawachi site, where celadon wares were made in great variety—as much as twenty-five per cent of the production 'seems to have been celadon of fine quality'.[1] A fine example of this Nabeshima celadon ware is shown in Plate 32; while a few Kakiemon celadons have also survived, some being figurines or in open-work—perhaps derived from Korean celadons in the same style.

[1] Soame Jenyns, *Japanese Porcelain*, London, 1965, p. 237.

Probably the most famous individual potter to have made celadon wares was Aoki Mokubei of Kyōto (1767–1833), a great artist, connoisseur and devotee of the Tea Ceremony. He is said to have made tea-bowls in the Korean tradition and to have excelled at copying Chinese celadon and blue-and-white. At about the same time celadon was also being made at the Kameyama kiln, near Nagasaki (active 1804–67), also at Arita, Hirado, Kutani and Sanda. Hobson noted that Chinese celadons were closely imitated in Japan, 'where the old Sung specimens were treated with an almost reverential respect', and stated that 'it was a specialty of the Sanda pottery established in the Arima district of Settsu in the nineteenth century. The material found in this district was peculiarly suitable for celadon ware, and the paste of the Sanda celadon has much similarity to that of the Chinese ware of Lung-ch'üan. There was also a factory at Himeji, in the province of Harima, where celadon and blue and white porcelains were made with materials from Tozan hill.'[1] A small series of these Japanese celadons may be viewed in the British Museum.

In recent times another Kyōto potter, Sōtarō Uno, made copies of Lung-ch'üan ware which are works of genius, superb celadons in their own right but so close to the originals that experts have often been confused. The types specially favoured in Japan have been the *Kinuta* (mallet-shaped) vases, three-legged incense burners and *tobi seiji*, or celadons with iron-brown spots as decoration. For an interesting discussion of these Japanese imitations of Chinese celadon the reader is referred to an extract from a letter written by Mr. W. W. Winkworth, one of the most knowledgeable collectors of Japanese porcelain in the West.[2]

---

[1] R. L. Hobson, *Handbook of the Pottery and Porcelain of the Far East*, British Museum, London, 1937, p. 167.
[2] Soame Jenyns, *op. cit.*, pp. 282–5.

CHAPTER SIX

Sawankalok Celadon

Accerding to Thai tradition, the pottery centres established some two hundred miles north of Bangkok were set up at the end of the thirteenth century by Chinese potters brought back by King Khamhêng from a mission to the Yüan court. The King led two missions to Peking, in 1294 and 1300, and the story goes that he was so impressed with the beauty of the Chinese wares he saw that he requested permission to take back with him some Chinese potters to assist in starting factories in Siam. It seems likely that two groups of potters came to Siam, the first settling at the capital a few miles west of the present town of Sukhothai and making products which were clearly modelled on Tz'ŭ-chou ware. The second group, which may have come some years later, settled on the Yom River, about sixty miles to the north, and here, from about the middle of the fourteenth century, an entirely different class of ware was produced which closely resembles the celadons made at the great Chinese centre of Lung-ch'üan in Chekiang Province.

Compared with the Sukhothai wares, Sawankalok celadon seems to represent the work of more accomplished craftsmen, and the variety and scale of production grew to the point where it became a staple export: indeed, it seems to have been produced primarily for this purpose, and it is therefore hardly surprising that some of the best collections nowadays are to be seen in the Philippine Islands and Indonesia rather than Siam itself. There are plain, undecorated pieces as well as others having designs incised or impressed beneath the glaze, and the shapes include large plates, pots, jars, bottles, vases and human or animal figures. The body varies from a light grey stoneware to a white semi-porcelain, both types burning a reddish colour where unglazed, and the glaze is a translucent, watery grey-green, with a tendency to run into pools at the bottom of bowls and dishes, where it becomes flecked or streaked with grey. Irregular crazing is a frequent feature, and the ware has a 'soft', rustic appearance—though it is hard and brittle to the touch—which greatly adds to its attractions. It is a plain, unostentatious ware, ideally suited for table ornaments 'as an adjunct to old furniture', to recall Hetherington's phrase (see page 23). Dr. Spinks, a leading authority on Thai ceramics, has ably summed up the impression it gives in the following words: 'The original qualities of clay and stone seem to be so fully preserved in some of the wares of Sukhothai and Sawankalok to suggest that they were truly born of the Earth, in consequence of which their seemingly crude workmanship and simple shapes and designs merely serve to emphasize their natural but enduring charm.'[1]

The decoration of Sawankalok celadon is in keeping with this general feeling of simplicity, consisting of engraved rosettes and petal bands, wheel-cut grooves cut on the shoulders of jars, channelled flutings on the exterior of bowls, with combed shading and roughly scored lines. A common form of bottle has a distinctive, globular body with two small loop-handles for a cord on either side of the short neck and mouth (Plate 35); another shape is a tall, elongated version of the same, usually seen as a miniature only a few inches in height and often having a bulbous swelling at the base of the neck. The glaze is subject to much variation as a result of inexpert technique: while normally thick and glassy, it sometimes shows bluish, opalescent effects or is pale green and opaque; a dense blue-grey or grey-green 'mutton-fat' type is also found.

The Sawankalok and Sukhothai kiln sites have been known since the 1880's, when they were discovered by French and English travellers; but the first detailed accounts date from the early years

[1] Charles Nelson Spinks, 'A Ceramic Interlude in Siam', *Artibus Asiae*, Vol. XXIII, 2, 1960, p, 105.

of the present century, when T. H. Lyle published articles and sent fragments back to England; but it was not until the 1930's that general interest was aroused by studies in the *Burlington Magazine*, the *Transactions of the Oriental Ceramic Society* and the Japanese journal, *Tōji*. A much-needed work on the subject has recently been published by Dr. C. N. Spinks (see Bibliography). At present the history remains somewhat obscure and dating is very uncertain, depending largely on analogies of style with the Chinese wares from which Sawankalok celadon was derived. A peculiarity of technique in the firing was that the vessels sometimes rested singly on tall, tubular supports varying in height from a few inches to nearly two feet; these left traces in the form of a dark ring on their bases. In other cases flat discs of fire-clay were employed with five short spurs on which bowls or dishes were stacked, so that spur-marks were left inside all but the top piece.

The Sawankalok wares are only comparable to Chekiang celadons of the second grade, and it would seem that the Chinese potters who came to Siam were by no means leading craftsmen, while their Thai apprentices could not even maintain the original standards—but nevertheless introduced native characteristics which lend their own special charm and distinction to the products of the kilns. It is evident also that the materials at hand for making clay and glaze were distinctly inferior to those which were available in Chekiang, while the construction of the kilns left much to be desired. Dr. Spinks states that 'kilns frequently collapsed during the firing, resulting in damaged or blemished pieces many of which had to be discarded as wasters and now constitute the great heaps of sherds and misshapen bowls found at the kiln sites'.[1] A considerable number of the wares also were misfired, with the result that they were warped or otherwise impaired.

The magnanimity of the Chinese in making available even the less skilled artisans from their potteries and thus extending to their tributaries some of the blessings of their own civilization made it

[1] Charles Nelson Spinks, *op. cit.*, p. 106.

possible for the Thais to build up what must have been a highly profitable industry. An inexhaustible demand arose in Southeast Asia, particularly in Borneo, Indonesia and the Philippines, for the new durable stoneware with vitreous glazes, which was vastly preferable to the local earthenware utensils made from prehistoric times. Not only were these celadons of much greater practical use for the serving and storing of food and water, but their beauty and mystery were highly appreciated in the realm of ritual and magic, and they fulfilled local needs for burial urns or other funerary purposes. Spirits were summoned at rites and festivals by the musical ring produced when the vessels were tapped; jars were also kept as talismans and to serve as the abode of spirits. Here again we see the widespread belief in the magical properties possessed by celadon wares, which seems to have extended from the Far East through central Asia to the Mediterranean and Europe. Indeed, Sawankalok celadon may well have met with an export success that disconcerted even the Chinese, who had been responsible for its development, for Dr. R. B. Fox, while excavating some fifteenth-century burial sites at Calatagan, about sixty miles south of Manila, discovered some contemporary Chinese bowls which had a glassy, light green glaze exactly like that of Sawankalok ware, and Professor Beyer has suggested that these may represent an attempt on the part of the Chinese to break into the market established by the Thais by imitating the typical Sawankalok glaze.[1]

However, Sawankalok celadon was produced for only a relatively short period. The ceramic industry seems to have come to an abrupt end at some point in the latter part of the fifteenth century, after a florescence of little more than a century. There is no reason to suppose that the demand fell off at this time, for the Ming celadons and blue-and-white wares from China were enjoying a period of great prosperity; the collapse must be rather

[1] Robert B. Fox, 'The Calatagan Excavations. Two 15th Century Burial Sites in Batangas, Philippines', *Philippine Studies*, Vol. 7, No. 3, August 1959, pp. 321–90.

attributed to the struggle going on at that time between the Thai states, which resulted in the town of Sawankalok being destroyed in 1460 by invading forces from Chiengmai. This was not the reason for Chinese celadons being exported to the Philippines, since a number of Sawankalok wares also were excavated at Calatagan, indicating that celadons were reaching the islands from both sources in the early fifteenth century. However, Sawankalok celadons are only rarely found at late fifteenth- or sixteenth-century sites, confirming that the export of these wares had virtually ceased by this time.

The sudden termination of the pottery industry at Sawankalok is proved by the discovery of kilns filled with unfinished wares. Evidently the potters either took to flight or were evicted at short notice, some of them apparently setting up at Chiengmai, where they endeavoured to carry on their trade. The rise of Chinese blue-and-white porcelain may also have been a contributory cause to the abandonment of the Thai industry and cessation of exports.

Of course the production and export of wares from Siam never approached that of China: it was a small-scale industry devoted almost exclusively to supplying the needs of the South-east Asian countries, which were satisfied with lower-grade wares than the Chinese. However, Dr. Spinks is justified in claiming for the Thai wares a position of real importance in the history of oriental ceramics. As he writes: 'it is impossible to confuse these wares with those of any other kilns: the potters of Sukhothai and Sawankalok achieved an enduring quality in the unsophisticated simplicity of their work that transcends whatever they may have lacked in the way of technical competence or artistic accomplishment. . . . The comparatively short-lived production of these unique stonewares (represents) a strange but brilliant episode in the cultural history of the Thai'.[1]

In recent years the kiln sites at Sawankalok have been systematically plundered and deep pits and trenches dug to recover as much

[1] Charles Nelson Spinks, *op. cit.*, p. 110.

as possible of the kiln waste. Nearly all the curio shops in Bangkok have small cases full of wasters, including bowls distorted in the firing or fused together as a result of mishaps. The prices of these relics have risen astronomically, and it is no longer possible—as it was until quite recently—to make an interesting collection of the wares for a modest outlay.

CHAPTER SEVEN

Modern Celadon, Copies and Fakes

M ost modern potters have tried their hand at making celadon
ware with good results, now that so much is known about
the scientific basis and the effect of firing in a reducing or oxidizing
atmosphere. Modern potters use chemically purified oxides, but
it is still feasible to employ natural materials, and this has been
done by many of them experimentally with various different
kinds of wood-ash for use in the glaze. Bernard Leach, whose
Potter's Book should be read by all who have any interest in collecting or the potter's art,[1] lists six types of celadon glaze, of which
two hark back to Chinese Lung-ch'üan ware, two to Korean
celadon, while two are purely Japanese. However, relatively few
modern potters have specialized in producing celadon ware. In
the early 'fifties James Walford was making a variety of celadons
based on the Chinese wares of the Sung period, some of which
were very successful, though in general more thickly potted than
the better Lung-ch'üan or Northern Celadon wares. At the
Crowan Pottery near Camborne in Cornwall, Harry and May
Davis for some years produced some of the best celadon ware for
table use, and this was on sale at the Crafts Centre in London and
Heffer's in Cambridge at very moderate prices. The same kind of
utilitarian celadon ware is produced by a number of Japanese
potters and may be seen at department-stores and pottery shops in
Tōkyō; but it does not seem to have caught the public fancy to the
same extent as the various oatmeal-cream, blue-grey and *temmoku*
type (glossy black or brown) wares made by the Leach Pottery at

[1] Bernard Leach, *A Potter's Book*, Faber & Faber, London, 1940. (Second
edition, 1945.)

93

St. Ives, the Hamada potteries at Mashiko in Japan and hosts of other studio-potters working in much the same style all over the world.

In fact, the only large-scale production of celadon ware at the present time seems to be that mentioned on the first page of this book, sold by the Thai Celadon Company, Ltd., at Bangkok. This is a highly successful and popular type based on the four-teenth- and fifteenth-century Sawankalok celadons. The kilns are situated at Chiengmai—where some of the Sawankalok potters are said to have settled after being forced to flee from their original potteries in 1460—and are fired with wood-fuel, like those of South China, Korea and Japan. Natural deposits of stone and feld-spar are found at the kiln site and the raw materials for the glaze are carried down from the jungle by mountain tribesmen, who prepare the ash used in the mixture from the bark and leaves of trees indigenous to the area. No synthetic materials or commercial clays or dyes are used, and each piece is thrown by hand and has the individual appeal which derives from the entirely natural processes and earthy materials employed. Complete table services are made as well as a large variety of more or less standard pieces for everyday use, such as beer mugs, tea-pots, leaf-shaped dishes, stem-cups, sweet dishes, bulb bowls, fruit platters, casseroles, water jars, vases and lamp-stands; also book-ends in the form of *naga* snake-dragons or Siamese *tepanon* guardian angels together with oriental masks and models of tigers, crocodiles, etc., as table orna-ments. The two principal glazes are the light green type similar to Sawankalok celadon and a darker olive-green called 'jungle green'. These Thai Celadon products are most attractive and not unduly expensive, but they are heavily potted and weighty pieces for the most part. The decorative designs are conventional: in-cised floral patterns or embossed work in relief. Many of the shapes are similar to those found in Sawankalok celadon, that is to say they are traditional and typically Siamese, with a distictive flavour quite different from Chinese or Korean wares.

Mention has been made of the copies produced by skilled

potters in Japan, and also in China, especially of Lung-ch'üan celadon. Dr. Ch'ên notes that some of these were made in Hopei Province at the time of the War with Japan, i.e. in the 1930's and early 1940's, and subsequently 'with great success' at Ching-tê chên. Some of them are diabolically clever and would deceive all but the most experienced, especially when they take the form of wasters, marred by some obvious defect and thus masquerading as kiln site finds. One ingenious example seen by the writer is a pair or more of mallet-shaped vases fused together apparently by some mishap in the firing and provided with an elegant carved wooden stand; however, most of the higher class Chinese products were fired in separate saggars, so that such an accident would be extremely unlikely. Copies have also been made of Korean celadons, but this presents unusual difficulty because, as Dr. Yanagi succinctly put it in conversation with the writer, 'the materials are special and the glaze very thin'; but this would not prevent Korean potters from producing similar wares, using the same local materials, and that in fact is what has been done. Before the Second World War excellent signed copies of Korean inlaid celadon were made at a pottery, doubtless under Japanese direction, in Seoul: while the decoration was actually inlaid, probably with a brush, the wares have such a finished and 'machine-made' appearance that they would not deceive any experienced collector; and indeed it was not their object to do so, for they were clearly marked as copies and were on sale at reasonably low prices in the department-stores in Seoul.

However, the whole subject of copies and fakes is extremely involved. Dealers are well aware how many dubious specimens find their way onto the market, and there is no sure guide to their detection apart from experience. Moreover, many of them may well be centuries old, so that an appearance of age is not necessarily indicative; and they may well have been made in the very region where the original famed celadon wares were produced in the twelfth century or later. The first thing that strikes an experienced dealer or collector about a clever copy is that there is something

'fishy' about it which cannot be defined at all clearly; close examination may or may not add to this impression, since the maker
will have been at pains to provide skilfully improvised 'clues'
such as one would expect to find in authentic pieces. But, with the
passage of time, the feeling that there is something 'wrong' about
the specimen is likely to become intensified; and then one would
be very ill-advised to take the chance and pay a high price for it.
When confronted with one of these doubtful specimens, the writer
is disinclined to pronounce it a fake, contenting himself with the
comment that it is not likely to be a twelfth century (or whatever
is the best period) ware but a later copy in the same style. It may
be only one hundred years later, or quite modern; but that is a
matter requiring more knowledge than anyone is likely to possess
at the present time.

The golden rule for the new collector is thus to familiarize himself with the principal wares, to read all he can about them in
books written by standard authorities, to study them in museums
and, if possible, private collections and, above all, to make arrange-
ments to *handle* them—not always easy in view of the many
claims on overworked museum curators. By no other means can
anyone gain the experience and 'feel' necessary to appraise any
type of porcelain; and finally, let no beginner expect to discover
bargains by visiting 'junk shops' and the like: the experience and
knowledge of a reputable dealer will always be available, and the
increased price which is necessary to obtain authenticated pieces
is well worth paying, if only as a protection against making
expensive mistakes.

The reason for the production of so many copies and fakes is, of
course, the increasing demand for specimens and the high prices
they command. So many of the genuine wares have perished in
wars, civil disturbances and the normal vicissitudes of fire and
flood, breakage and accident. The beauty and rarity of authentic
examples seem to grow with the passing of time, but we can be
thankful that it is still possible to enjoy the experience of seeing
these wonderful examples of the potter's art at so many museums

D. Deep bowl with cover. Siamese. Sawankalok ware. Height $9\frac{3}{4}$ in. Diameter $7\frac{1}{4}$ in. British Museum.

and in private collections in the West as well as in the countries where they were made. For this reason we can rejoice that the misgivings expressed by Kao Lien in his *Tsun-shêng Pa-chien* in 1591 can be appreciated without being realized. He wrote:

'It is impossible to foretell to what point the loss of these ancient wares will continue. For that reason I never see a specimen but that my heart dilates and my eye flashes, while my soul seems suddenly to gain wings and I need no earthly food, reaching a state of exaltation such as one could scarcely expect a mere hobby to produce. My great grief is the thought that those who come after me will hear the names of these wares, but never see the wares that bore these names'.[1]

[1] As translated by Arthur Waley, *op. cit.*

Short Bibliography

IN ENGLISH

John Ayers, *The Seligman Collection of Oriental Art, Vol. II: Chinese and Korean Pottery and Porcelain*, London, 1964.

S. W. Bushell, *Description of Chinese Pottery and Porcelain, being a Translation of the T'ao Shuo*, Oxford, 1910.

G. St. G. M. Gompertz, *Chinese Celadon Wares*, London, 1958.

G. St. G. M. Gompertz, *Korean Celadon and Other Wares of the Koryŏ Period*, London, 1963.

Basil Gray, *Early Chinese Pottery and Porcelain*, London, 1953.

R. L. Hobson, *A Catalogue of the Chinese Pottery and Porcelain in the Collection of Sir Percival David*, London, 1934.

R. L. Hobson, *Chinese Pottery and Porcelain*, 2 vols., London, 1915.

W. B. Honey, *The Ceramic Art of China and Other Countries of the Far East*, London, 1945.

W. B. Honey, *Corean Pottery*, London, 1947.

Chewon Kim & G. St. G. M. Gompertz, *The Ceramic Art of Korea*, London, 1961.

Nils Palmgren, W. Steger, N. Sundius, *Sung Sherds*, Stockholm, 1963.

G. R. Sayer, *Ching-tê-chên T'ao Lu, or the Potteries of China*, London, 1949.

Charles Nelson Spinks, *The Ceramic Wares of Siam*, Bangkok, 1965.

Charles Nelson Spinks, 'Siam and the Pottery Trade of Asia', *The Journal of the Siam Society*, Vol. XLIV, Pt. 2, Bangkok, 1956.

Michael Sullivan, *Chinese Ceramics, Bronzes and Jades in the Collection of Sir Alan and Lady Barlow*, London, 1963.

Lorraine d'O. Warner, 'Kōrai Celadon in America', *Eastern Art: An Annual*, Vol. II, 1930.

William Willetts, *Foundations of Chinese Art*, London, 1965.

SHORT BIBLIOGRAPHY

The Chinese Exhibition: A Commemorative Catalogue of the International Exhibition of Chinese Art, Royal Academy of Arts, Nov. 1935—March 1936, London, 1936.

Porcelain of the National Palace Museum:

Ju Ware of the Sung Dynasty
Kuan Ware of the Sung Dynasty
Kuan Ware of the Southern Sung Dynasty:-
  Book 1. Part 1: *Hsui-Nei-Ssŭ Kuan Ware*
  Book 1. Part 2: *Hsui-Nei-Ssŭ Kuan Ware*
  Book 2. *Chiao-T'an-Hsia Kuan Ware*
  Chün Ware of the Sung Dynasty
  Lung-Ch'üan Ware of the Sung Dynasty

National Palace Museum and National Central Museum, Taichung, Taiwan, 1961–2.

In JAPANESE

Fujio Koyama, *Shina Seiji Shi-kō* (Sketch History of Chinese Celadon), Tōkyō, 1943.

In CHINESE

Wan-li Ch'ên, *Chung-kou Ch'ing-tz'u Shih-lüeh* (Outline History of Chinese Celadon), Shanghai, 1956.

Wan-li Ch'ên, *Yüeh Ch'i T'u-lu* (Illustrations of Yüeh Ware), Shanghai, 1937.

Index

INDEX

INDEX

INDEX

INDEX

1. Ewer with 'chicken-head' spout. Yüeh ware. Six Dynasties. Ht. 18½ in. Hakone Art Museum (Japan).

2. Basin with incised decoration of two fishes and wave patterns.
Yüeh ware, probably made at Chiu-yen. Han dynasty.
Diam. 13¾ in. Museum of Eastern Art, Oxford. (Ingram Collection).

3. Jar decorated with spots of iron-brown. Yüeh ware.
Six Dynasties. Ht. 8½ in. Japanese Collection.

4A. Small jar decorated with iron-brown spots. Yüeh ware.
Six Dynasties. Japanese Collection.

4B. Lion candlestick or water vessel. Yüeh ware, probably made at
Chiu-yen. Six Dynasties. Length 5 in. Museum of Fine Arts, Boston
(Hoyt Collection).

5A. Jar with four spouts and incised peony on the side. Yüeh ware, probably made at Shang-lin Hu. Five Dynasties. Japanese Collection.

5B. Celadon shards with finely incised decoration from Shang-lin Hu. Five Dynasties. Mr. Junkichi Mayuyama.

6. Ewer with carved floral decoration and loop in flower form on
either side of neck. Possibly Li-shui ware. Sung dynasty.
Ht. 6½ in. Barlow Collection.

7. Jar with four handles. Northern ware. Sui dynasty. Ht. 7½ in.
Victoria and Albert Museum. Crown Copyright.

8. Dish. Ju ware. Sung dynasty. Diam. 5·1 in.
Percival David Foundation.

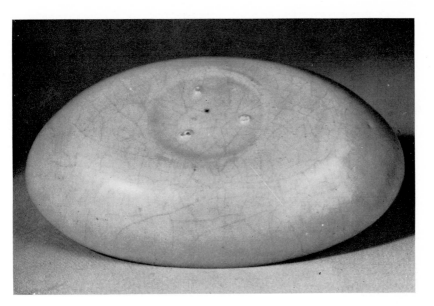

9. Elliptical brush-washer (under-side). Ju ware. Sung dynasty.
Length 5·6 in. Percival David Foundation.

10. Ewer with strongly carved floral decoration under a pale green
glaze. Possibly Tung ware. Sung dynasty. Ht. 8½ in.
Cunliffe Collection.

11. Dish with carved peony decoration. Northern Celadon.
Sung dynasty. Diam. 7¾ in. Barlow Collection.

12. Octagonal bottle with long tapering neck ending in a flange;
horizontal ribs on the neck and body. Greyish blue glaze.
Southern Kuan ware. Sung dynasty. Ht. 8·7 in.
Private Collection.

13. Dish in the form of an eight-petalled flower covered with a
bluish glaze. Southern Kuan ware. Sung dynasty. Diam. 6·7 in.
Percival David Foundation.

14A. (*top left*) Bottle with globular body and long neck.
Bluish green glaze. Southern Kuan ware (?). Sung dynasty.
Ht. 9·1 in. Japanese Collection.

14B. (*bottom left*) Brush-washer with straight sides and wide flat rim.
Bluish green glaze. Southern Kuan ware (?). Sung dynasty.
Diam. 5·7 in. Mr. and Mrs. F. Brodie Lodge.

15. (*above*) Dish with pair of fishes in relief. Lungch'üan
ware. Sung dynasty. Diam. 8·2 in.
Mr. Desmond Gure.

16. Mallet-shaped vase with phoenix handles. Lung-ch'üan
ware (Kinuta type). Sung dynasty. Ht. 10·4 in.
Yōmei Bunko Collection.

17A. Vase with flaring mouth and floral decoration in relief. Lung-ch'üan ware. Sung dynasty. Japanese Collection.

17B. Bowl with moulded petal decoration outside. Lung-ch'üan ware. Sung dynasty. Diam. 9¼ in. Col. J. G. Figgess.

18. Vase of spotted celadon ('tobi seiji'). Lung-ch'üan ware.
Sung dynasty. Ht. 10¾ in. Victoria and Albert Museum.
Crown Copyright.

19. Basin with dragon, cloud scrolls and flowers in biscuit relief.
Lung-ch'üan ware. Yüan dynasty. Diam. 16·95 in.
Percival David Foundation.

20. (*above*) Beaker-shaped vase in the form of a bronze 'tsun'.
Lung-ch'üan ware. Ming dynasty. Japanese Collection.

21. (*right*) Dish with incised floral design; on the base the inscription
'Great Ming Hsüan-te Lung-ch'üan.' Lung-ch'üan ware.
Probably Ch'ing dynasty. National Museum, Tokyo.

22. Incense burner with cover surmounted by a lion. Korean ware. Koryŏ dynasty. Ht. 8·4 in. National Museum of Korea.

23. Lobed vase of elongated melon shape with foliate
mouth. Korean ware. Koryŏ dynasty. Ht. 9¾ in.
National Museum, Tokyo.

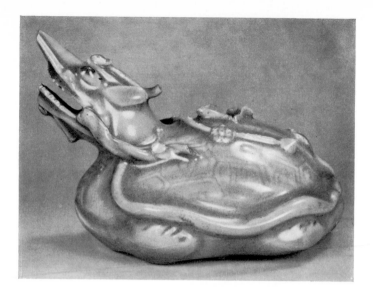

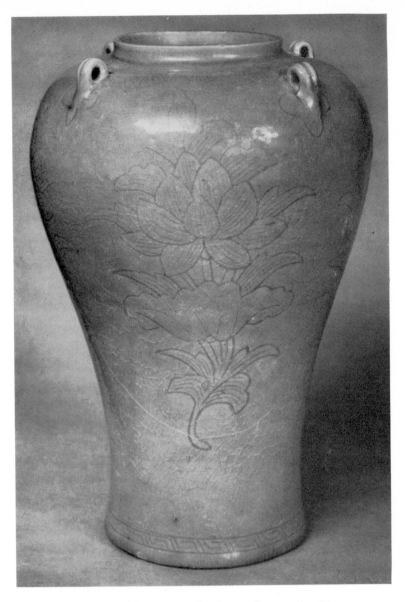

24A. (*top left*) Water-dropper in the form of a tortoise. Korean ware.
Koryŏ dynasty. Ht. 3½ in. Japanese Collection.

24B. (*bottom left*) Incense burner in the form of a bronze 'ting' with
moulded decoration. Korean ware. Koryŏ dynasty. Ht. 6¾ in.
Japanese Collection.

25. (*above*) Jar with loop handles and incised lotus decoration. Korean
ware. Koryŏ dynasty. Ht. 10 in. Author's Collection.

26. Bottle with incised lotus decoration. Korean ware.
Koryŏ dynasty. Ht. 12⅝ in. Author's Collection.

27. Wine pot and bowl with incised lotus decoration. Korean ware.
Koryŏ dynasty. Ht. 7½ in. Author's Collection.

28. Ritual ewer or 'kundikā' with incised decoration of lotuses,
bamboos, cranes and clouds, ducks swimming and flying,
chrysanthemums, etc. Korean ware. Koryŏ period.
Ht. 13½ in. Author's Collection.

29. Wine cup and stand inlaid with chrysanthemums. Korean
ware. Koryŏ dynasty. Ht. 5⅛ in. Author's Collection.

30A. Small pot with cover and saucer decorated with
floral scrolls inlaid by the 'reverse method',
i.e. background inlaid. Korean ware.
Ht. 2½ in. Koyrŏ dynasty.

30B. Small box and cover decorated with inlaid peonies
and chrysanthemums. Korean ware. Koyrŏ dynasty.
Diam. 3·4 in.

Author's Collection.

31. Japanese Celadon: Seto Ware: first quarter of the XIVth century, with incised floral decoration. Ht. 10½ in. Japanese Collection.

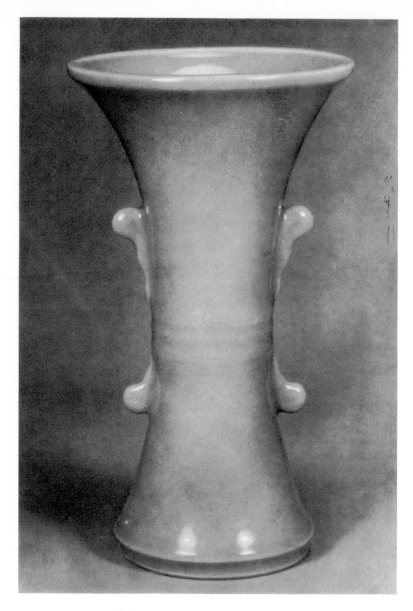

32. Japanese Celadon: Vase of Nabeshima ware. Mid-eighteenth century. Ht. 9·4 in. Japanese Collection.

33. Japanese Celadon: Bowl with floral decoration in relief.
Arita ware. Eighteenth century. Ht. 3 in. Length $6\frac{3}{8}$ in.
Width 5 in. Mr. Richard de la Mare.

34A. Japanese Celadon: Covered box with floral
decoration in relief surmounted by a modelled
lion; made by Yaichi Kusube in 1952. Ht. 6 in.

34B. Japanese Celadon: Incense burner with floral decoration
in relief; made by Yaichi Kusube in 1939. Diam. 9 in.

35. Globular bottle with two small loop-handles. Sawankalok ware. Ht. 6¼ in. Victoria and Albert Museum. Crown Copyright.

36A. Small Sawankalok Celadon bottle with ears. Ht. 4⅝ in. Victoria and Albert Museum. Crown Copyright.

36B. Modern Celadon wares produced by the Thai Celadon Company of Bangkok. Ht. of bottle 4¼ in.